D1286171

On the Shoulders of Medicine's Giants

Robert B. Taylor

On the Shoulders of Medicine's Giants

What Today's Clinicians Can Learn
from Yesterday's Wisdom

 Springer

Robert B. Taylor
Department of Family Medicine
Oregon Health and Sciences University
School of Medicine
Portland, OR, USA

Department of Family and Community Medicine
Eastern Virginia Medical School
Norfolk, VA, USA

ISBN 978-1-4939-1334-3   ISBN 978-1-4939-1335-0 (eBook)
DOI 10.1007/978-1-4939-1335-0
Springer New York Heidelberg Dordrecht London

Library of Congress Control Number: 2014954588

Printed on acid-free paper

Springer is part of Springer Science+Business Media (www.springer.com)

In the continual remembrance of a glorious past individuals and nations find their noblest inspiration.

—Sir William Osler (1849–1919)

The leaven of science. Philadelphia: Wister Institute of Anatomy and Biology of the University of Pennsylvania; 1894, page 79.

Medicine is the oldest learned profession in the world and it is rooted in the past. Each successive generation of doctors stands, as it were, upon the shoulders of its predecessors, and the fair perspectives that are now opening before you are largely the creation of those who have gone before you. It is therefore reasonable to think that anyone who has spent a long professional life in medicine must have something to pass on—however small or modest.

—British neurologist Sir F.M.R. Walshe (1888–1973).

In: Canadian Medical Association Journal. 1952;385:67.

You have chosen the most fascinating and dynamic profession there is, a profession with the highest potential for greatness, since the physician's daily work is wrapped up in the subtle web of history. Your labors are linked with those of your colleagues who preceded you in history, and those who are now working all over the world. It is this spiritual unity with our colleagues of all periods and all countries that has made medicine so universal and eternal.

—Spanish-born American physician and educator Felix Martí-Ibáñez (1911–1972).
Epilogue: To be a doctor.

In: A prelude to medical history. New York: MD Publications; 1961, page 197.

Preface

This book, intended to make you a more knowledgeable clinician, presents selected insights of some of the history's leading physicians, scientists, and scholars—the admonitions of Hippocrates, what Edward Jenner had to say about the end of smallpox, Sir William Osler's thoughts about uncertainty and probability in medicine, and Florence Rena Sabin's vocal commitment to believing in her work—and describes how their words are pertinent to the current practice of medicine. This book will show how much the thoughts of medicine's giants were prescient, and are manifested in what we believe and do today.

The book's title is inspired by the words of the English physicist and mathematician Isaac Newton in 1676: "If I have seen further it is by standing on the shoulders of giants" [1]. There is, of course, evidence that the metaphor describing *standing on shoulders* of predecessors predated Newton in various iterations [2]. Attribution squabbles notwithstanding, as we ponder this image, an example that comes to mind is the 1854 prophetic comment attributed to John Snow that the key to elimination of the great plagues, such as cholera, would lie in understanding how they are propagated. Today's knowledge of disease prevention is built upon the findings of Snow and others like him.

In this world of electronic medical records, virtual physicians, and *nocturnalists* (physicians who choose to work the "graveyard" shift in hospitals), why do we need a book based on the wisdom of our forebears? Perhaps the answer lies in the premise of the query: We have become deeply enchanted with the twenty-first century's dazzling technology, expansive communication options, and lifestyle-oriented practice choices. And in doing so, we risk losing touch with the passion for clinical excellence and commitment to service that made medicine what it is today.

There is another reason this book is important. Today's healers enjoy a level of societal confidence and trust that is the envy of our colleagues in other professions. In the 1940s, American author and newspaper columnist Damon Runyon quipped: "My old man used to say that he guessed the percentage of scoundrels was less among doctors than any other class of men, professional or otherwise, in the world" [3]. This lofty status exists only partly because of the good works of we, the living, but is much more a legacy of the dedication, perseverance, and stature of those who created today's house of medicine—the medical giants upon whose shoulders we stand. If only for this reason, we should all spend some time learning about our heritage.

The topics discussed are diverse. They range from basic science to philosophy, from doctors to patients and their families, and from classic descriptions of disease to how clinical caregivers view their world. Some quotations, such as Lewis Thomas' observation that "most things get better by themselves" are presented with a hint of irony as we consider the heroic therapy often employed in modern clinical practice. Other discussions, such as Albert Schweitzer's commitment to service, are unashamedly inspirational, and some have a touch of pragmatism, as in William Heberden's advice to cease doctoring at the right time, before one can no longer do justice to patients. Some are reflective, as the thoughts of Elizabeth Blackwell, America's first female medical school graduate, about being a pioneer. And I include a few notions that time has proved to be quite erroneous.

Is this just another book of medical quotations? Not really. The thoughts presented are, in most cases, whole paragraphs, allowing greater elucidation of the authors' ideas than is possible with a single adage. One example is the paragraph taken from the 1927 article *The Care of the Patient*, by Francis W. Peabody, although, as you will see in Chap. 3, I might have quoted only the memorable last sentence. In the case of Elizabeth Kübler-Ross M.D., the selection is both extended and metaphorically vivid. Robert Lewis Stevenson's representation of physicians as standing "above the common herd" cannot be summarized in a few words. On the other hand, pithy quotes such as the observation by Sir Dominic Corrigan that "The trouble with doctors is not that they don't know enough, but that they don't see enough" might reasonably be termed aphorisms. There are even a few poems included in the book.

Are all quotations from the past? Not by any means. In searching for sayings that resonated with my idea of what medicine should be, the characteristics of the ideal physician, and the broad panorama of medical practice in history and today, I came across a number of thoughtful insights penned by "modern" physicians, scientists, and other writers. And while some may hold that these persons have not yet been accorded the status of "giant," I felt that their words merited inclusion in the book. Thus, in addition to Celsus, Hunter, and Pasteur, I have included the writings of some individuals, such as Edmund Pellegrino and Barbara Starfield, who have shared our time on this earth, and others, including Abraham Verghese and David Hilfiker, who are still contributing to medicine.

In assembling my sources, I was forced to make some decisions. First of all, should I return to the same source more than once? In fact, I could easily have presented an Oslerism on every third page. I decided that, for the sake of variety, I would cite major extracts from the works of Galen, Freud, Sydenham, and other familiar names only once in a chapter, and that I would lean toward including a broad range of "giants." Hippocrates, Virchow, Osler, and a few others do, however, show up more than once in the book.

I also faced the question: Who is medical giant? It is someone, and not necessarily a physician, whose thoughts, words, and deeds have helped make medicine what it is today. I concluded that some writings by nonmedical authors such as Aristotle, Rudyard Kipling, and Michael de Montaigne have influenced how we view medicine and physicians, and hence their words are included.

A few giants have given us a treasury of medical quotes, sometimes including a notable "signature" adage. In these instances, I have given preference to the well-known saying, even though some writings that are less well known are also perceptive. For example, the words "Chance only favors the prepared mind" will always be linked to Louis Pasteur.

A more vexing issue was this: What about publications with multiple authors? I have found some astute thoughts in committee-authored articles, some with as many as five or six contributors. I decided that these words had been polished by many hands and also, I suspect, were subsequently buffed by unnamed editorial assistants. Thus I could not describe the entire team of authors as "giants," although some multiauthor publications are cited in my annotated comments.

For the most part, this book is grounded in literature, not simply the product of my personal reflections (although there are more than a few of the latter), and source-based annotations are supported by reference citations. Most are from the mainstream medical literature—published books, historical documents, peer-reviewed journals such as *The New England Journal of Medicine*, and so forth. And, in an effort to make the content timely and relevant, I have also used some less traditional resources such as the *Wall Street Journal*, the *New York Times*, and the *World Wide Web*.

This book may never lead you to a "eureka" diagnosis or guide a brilliant therapeutic drug choice. I hope that it will, however, influence how you think about your profession, your patients, and your career, and help you avoid the fate of becoming an "automated medical kiosk" [4]. I will be delighted if this book helps rekindle your idealism regarding what medicine is at its core, and that it reminds you how being a physician can be what Osler termed "a daily joy" [5].

On the Shoulders of Medicine's Giants is intended not as a text or reference source, but as an *enrichment book*. It is, fundamentally, a collection of perceptive quotations, with comments. Perhaps you will experience an "Aha" moment of recognizing the historical origins of why we think this or do that. I hope that you will find it fun to read—perhaps on a quiet evening or while passing the time on a long plane ride—and that your life will be a little richer for it.

1. Turnbull HW, Ed. The correspondence of Isaac Newton, vol 1. London: Cambridge University Press; 1959; page 416.
2. Dorizzi RM. Standing on the shoulders of giants—Isaac Newton? Bernard of Chartres? Priscian! (letter). Pharos. Alpha Omega Alpha Honor Med Soc. 2012;75:1.

3. Runyon D. From the author's long-running newspaper column: "The Brighter Side." c.1943. Available at: http://nostrums.blogspot.com/2010_03_01_archive.html.
4. Bynum W. Why physicians need to be more than automated medical kiosks. Acad Med. 2014;89:212.
5. Osler W. Aequanimitas, ed 3. Philadelphia: Blakiston; 1932, page 423.

Portland, OR, USA   Robert B. Taylor, M.D.
Norfolk, VA, USA

Contents

Chapter 1
The Profession and Professionalism

At the heart of medicine is its tradition as a healing profession. Through the millennia from shamans to today's robotic-enabled surgeons, our role as healers has bound us together. We have been *professionals*. In the classic sense, we have "professed" an oath, whether the words are attributed to ancient Chinese or Indians, Hippocrates, Maimonides, or, more recently, the Declaration of Geneva. On a more practical note, we have learned, employed, and shared a specialized body of knowledge and skills, and have been granted special status in society and exclusive rights of practice, i.e., licensing. We serve, but we also enjoy extraordinary privileges.

Fundamental to the belief system of medical practitioners is the philosophy that what we do is, first and foremost, for the benefit of humankind—often represented by the patient before us in the office or hospital. The patient comes first, ahead of economic gain or personal career advancement. This commitment to altruistic values is the cornerstone of medical professionalism. It is why the sick person can visit a new physician with the assumption that he or she will receive not only skilled, but principled care. This selfless commitment to caring for patients is a fundamental concept in professionalism.

Over the last decade or two I have read a lot about professionalism—generally defined as denoting the attributes, competence, and skill of being a "professional," as described briefly above. Not all "professionals" can claim to exhibit professionalism: As extreme examples I cite priests who sexually molest children, lawyers who cheat their clients, and physicians who fraudulently bill for services they did not perform.

© Springer Science+Business Media New York 2015
R.B. Taylor, *On the Shoulders of Medicine's Giants*,
DOI 10.1007/978-1-4939-1335-0_1

Today professionalism is a required competency in graduate medical education, as well as for board certification and maintenance of certification [1]. If professionalism is required, then how is it defined in a way that we can measure whether one does or does not possess the attributes?

In 2002, the *Annals of Internal Medicine* and *The Lancet* simultaneously published "Medical Professionalism in the New Millennium: A Physician Charter," containing three principles and ten commitments. The three principles were based on the values of traditional medical ethics: the primacy of patient welfare, patient autonomy, and social justice. The ten commitments range from clinical competence to appropriate relations with patients to professional responsibilities [2].

Then in 2009, we were presented with "A Blueprint to Assess Professionalism: Results of a Systematic Review." This report identifies five "clusters of professionalism": adherence to ethical practice principles; effective interactions with patients and with people who are important to those patients; effective interactions with people working within the heath system; reliability; and commitment to autonomous maintenance/improvement of competence in oneself, others, and systems [3].

And when it comes to professionalism as a required competency for physicians, the "Blueprint" provides guidelines for assessment: "a combination of observed clinical encounters, multisource feedback, patients' opinions, paper-based tests or simulations, measures of research and/or teaching activities, and scrutiny of self-assessments compared with assessments of others" [3].

As I read the papers cited above and others about professionalism, I am reminded of a quote (originally referring to *being in power*) by former British Prime Minister Margaret Thatcher. To paraphrase her analogy: Being professional is like being a lady; if you have to tell them you are, you aren't. But such an intuitive assessment is too simplistic for today's world. In fact, some of our assumptions about medical professionalism are evolving, as we will see in this and the following chapters.

1. Cohen JJ. Professionalism in medical education, an American perspective: from evidence to accountability. Med Educ. 2006;40:607.
2. Medical professionalism in the new millennium: a physician charter. Ann Intern Med. 2002;136:242.
3. Wilkinson TJ et al. A blueprint to assess professionalism: results of a systematic review. Acad Med. 2009;84:551.

Science, Art, and the Profession of Medicine

Life is short, science is long; opportunity is elusive; experiment is dangerous, judgment is difficult. It is not enough for the physician to do what is necessary, but the patient and attendants must do their part as well, and circumstances must be favorable.

Hippocrates (ca. 460–377 BCE) [1]

Of all the many writings attributed to Hippocrates, the thought recorded above is one of the best known (see Fig. 1.1). I find several intriguing facets to these two sentences. First of all, in checking several sources, I find the phrase "science is long" in some and "the art is long" in others. Today, of course, in the era of evidence-based medicine, we make a fairly clear distinction. Just as an example, a classic

Fig. 1.1 Hippocrates examining a child, a painting by Robert Thom, 1950s. http://www.spring-erimages.com/Images/MedicineAndPublicHealth/5-10.1186_1748-7161-4-6-11

textbook of diagnosis is titled *Sapira's Art and Science of Diagnosis* [2]. In medicine today, we think of science as what we can study and measure, and "the art" as based on experience and intuition. Has our thinking changed over the past 2,500 years?

The Hippocratic approach to medicine involved careful observation, natural healing methods, thoughtful prognostication, and avoiding harm to the patient, with the physician serving as nature's facilitator. Should we term such an approach "science" or "art?" Today's emphasis is certainly more on etiologic diagnosis and targeted intervention than was the style in the Age of Pericles. Hippocrates treated fever with "the drink made from coarse barley meal, sometimes from apple and pomegranate juice and juice from toasted lentils, cold" [3]. Today fever may prompt an expensive battery of laboratory tests, perhaps empiric antibiotic therapy or even, if the patient is an infant, a spinal tap.

Hippocrates believed, "Our natures are the physicians of our diseases" [1]. Hence the coda above that calls for nature to be aided by the actions of the patient and attendants—those who support the patient's quest for health—and favorable circumstances, such as a healthy lifestyle and the avoidance of environmental hazards. Could it be that Hippocrates and his contemporaries would hold that, in the practice of medicine, science and the art are actually closely related, maybe even one and the same? And to what degree are they linked today?

1. Lloyd GEF, Ed. Hippocratic writings. Translated by Chadwick J et al. New York: Penguin; 1983.
2. Orient JM. Sapira's art and science of bedside diagnosis, ed. 4. New York: Williams and Wilkins; 2009.
3. Hippocrates: Epidemics 2, 4–7, trans. Smith W. New York: Loeb; 1994.

Professionalism and Personal Gain

> No physician, in so far as he is a physician, considers his own good in what he prescribes, but the good of his patient; for the true physician is also a ruler having the human body as a subject, and is not a mere moneymaker.
>
> Plato (c. 429–347) [1]

Plato, student of Socrates and mentor to Aristotle, wrote on a wide range of topics, including logic, mathematics, ethics, religion, and, significantly for twenty-first century physicians, medical professionalism (see Fig. 1.2).

Even in ancient Greece, we encounter one of the ethical threats to professionalism—the physician focusing on personal gain ahead of the patient's well-being. Transgressions range from quackery to fraud to, well, simply

Fig. 1.2 Plato and Aristotle. A detail in Raphael's "School of Athens." http://www. springerimages.com/Images/ Education/1-10.1007_ s11858-009-0207-3-1

inappropriate professional money-seeking actions. This issue continues to be an existential threat to medical professionalism. Here is a random walk through the last few centuries of history:

- Physician Franz Anton Mesmer, in the eighteenth century, declared that he possessed "animal magnetism" and earned substantial fees using his supposed powers to treat the diseases of the rich and famous in Vienna and Paris.
- In the 1880s William Radam sold his remedy called *Microbe Killer*, touted to "cure all diseases." The solution, a mixture of dilute sulfuric acid and red wine, certainly would have annihilated microbes if applied directly. Taken internally, its use was potentially toxic.
- More recently, according to Rodin, California physicians accepted payments of $70 per patient for referrals to a hospital in Pasadena [2].
- The same source tells of physicians receiving 1,000 American Airline frequent flyer miles for each patient started on a certain brand of propranolol [2].
- A few years ago, a medical student reported to me that the physician who was teaching her in his office was selling Amway products in the waiting room. And a recent issue of *Dermatology Clinics* has an article telling how to sell skin care products in your "medspa" [3].
- In Dearborn, Michigan, a pediatric neurologist was accused of diagnosing healthy patients as epileptic to increase his income from tests performed [4].

Since 2007, the US federal government Medicare Fraud Strike Force, a multiagency group of investigators, has charged more than 1,400 defendants who collectively falsely billed the Medicare program more than $4.8 billion [5]. Yes, Plato, we have a dollar-denominated professionalism problem in medicine today.

1. Plato. Republic I.405 D.
2. Rodin M. Money, medicine, and morals. New York: Oxford;1996.
3. Mulholland RS. Selling skin care products in your medspa. Dermatol Clin. 2008;26:375.
4. 15 Fraud and abuse cases making headlines in 2010. Becker's Hospital Review. Available at: http://www.beckershospitalreview.com/hospital-management-administration/15-fraud-and-abuse-cases-making-headlines-in-2010.html.
5. Fields R. Health Care Fraud Prevention and Enforcement Action Team (HEAT) report: Available at: http://www.stopmedicarefraud.gov/aboutfraud/heattaskforce/index.html.

A Physician's Vow

This is my vow: To perfect my medical art and never to swerve from it so long as God grants me my office, and to oppose all false medicine and teachings. Then, to love the sick, each and all of them, more than if my own body were at stake. Not to judge anything superficially, but by symptoms, not to administer any medicine without understanding, nor to collect any money without earning it. Not to trust any apothecary, nor to do violence to any child. Not to guess, but to know.

Paracelsus (c. 1493–1541) [1]

Born a year after Columbus discovered America, Philippus Aureolus Theophrastus Bombastus von Hehenheim took the name Paracelsus, connoting that he was just as learned as the second-century Greek philosopher Celsus, author of *De Medicina*, regarded as a valuable record of ancient medicine. We will come to the words of Celsus in Chap. 6 (see Fig. 1.3).

Fig. 1.3 Paracelsus (Public domain). http://commons.wikimedia.org/wiki/File:Paracelsus.jpg

By all accounts, Paracelsus was an arrogant rascal, so much so that it has been alleged—incorrectly—that the word "bombastic" is derived from a word in his long original name. He was a pioneer in the use of chemical products in therapy, and his works included development of laudanum (tincture of opium), which has its own rich history. I point out the personal characteristics and his achievements as a preface to examining his writing cited above.

The first thing that strikes me about the paragraph quoted above is the title: *Jus Jurnadum*. Translated, this means a "sworn oath," as would be required in those days of a priest. Apart from the initial hubristic vow "to perfect my medical art" and the closing assumption that he can actually "know" and not be humbly uncertain at times, I am struck by the similarity in tone to the Hippocratic oath. I don't know if Paracelsus was familiar with the *jus jurnadum* (i.e., the oath) of Hippocrates; he certainly was the beneficiary of almost two millennia of medical progress.

Today we have updated the Hippocratic oath, with its transformation to the World Medical Association's *Declaration of Geneva* [2]. Might the current *Declaration of Geneva* be improved by adding some of Paracelsus' thoughts? I think about his words about false medicine and teachings, not judging things superficially, not using medicines we don't understand, and not collecting money we have not earned. Aren't these values all physicians should espouse today?

1. Paracelsus. Sketches, notes, and revisions: *Jus jurandum.* In: Paracelsus selected writings. Jacobi J, ed. Princeton NJ: Princeton University Press; 1995.
2. World Medical Association Declaration of Geneva. Available at: http://www.wma.net/en/30publications/10policies/g1/.

The Joy of Medicine

To each one of you the practice of medicine will be very much as you make it—to one a worry, a care, a perpetual annoyance; to another, a daily joy and a life of as much happiness and usefulness as can well fall to the lot of man.

Sir William Osler (1849–1919) [1]

Osler, described by Wheeler as "the most knowledgeable and competent clinician of his day," held a realistic view of the range of career satisfaction experienced by practicing physicians [2] (see Fig. 1.4). Simply stated, while many consider the practice of medicine a magnificent calling, not all are enthralled with their careers. A study of physician satisfaction across specialties revealed, using one of those familiar Likert-type scales, that despite all the hard work and current annoyances related to government and insurance company meddling in health

Fig. 1.4 Sir William Osler, ca. 1900. http://www. springerimages.com/Images/ MedicineAndPublicHealth/ 1-10.1007_s11845-007-0091-1-0

care, 70 % of US physicians described their career satisfaction as "satisfied" or "very satisfied."

Good news so far. But one in five reported that they were dissatisfied with their careers [3]. Doctors more satisfied with their practices often describe two variables: higher incomes and working in a medical school. What about the less satisfied physicians? They describe associations that include long work weeks, uncontrollable lifestyle, being a full owner of a practice, and reliance on managed care revenue [4]. All of this is important, of course, because dissatisfied physicians are "two to three times more likely to leave medicine than satisfied physicians" [5].

Osler's mastery of the science and art of medicine was built on a foundation of *service* to his patients. Once, when he was a young and impecunious physician, he gave his overcoat to a needy patient, even though lacking funds to buy another [1]. Today's medical school applicants are often asked, "Why do you want to be a doctor?" The usual answer involves some variation of *service to others*, often supported by touching personal examples. "High income" is a rare, and probably fatal, response. Years later, when a sense of career dissatisfaction looms, perhaps physicians should revisit the reason why they chose medicine as a career in the first place, and learn to find joy in, to use Osler's word, usefulness.

1. Osler W. Aequanimitas. Philadelphia: Blakiston; 1906.
2. Wheeler HB. Healing and heroism. N Engl J Med 1990; 322:1540.
3. Leigh JP et al. Physician career satisfaction across specialties. Arch Int Med. 2002;162:1577.
4. Leigh JP et al. Physician career satisfaction within specialties. BMC Health Services Research. 2009;9:166.
5. Landon BE. Leaving medicine: the consequences of physician dissatisfaction. Med Care. 2006;44:234.

The Privilege and Power of the Physician

You belong to the privileged classes. May I remind you of some of your privilege? You and kings are the only people whose explanation the police will accept if you exceed the legal limit in your car. On presentation of your visiting card you can pass through the most turbulent crowd unmolested and even with applause. If you fly a yellow flag over a centre of population you can turn it into a desert. If you choose to fly a Red Cross flag over a desert you can turn it into a centre of population towards which, as I have seen, men will crawl on hands and knees. You can forbid any ship to enter any port in the world. If you think it is necessary to the success of any operation in which you are interested, you can stop a 20,000-ton liner with mails in mid-ocean till the operation is concluded. You can order whole quarters of a city to be pulled down or burnt up; and you can trust to the armed cooperation of the nearest troops to see that your prescriptions are carried out.

British author Rudyard Kipling (1865–1936) [1]

Rudyard Kipling, one of the all-time favorite writers in a nation famous for its literary excellence, was respectful of the physicians in his life. In his treatise on *A Doctor's Work*, he wrote: "It may not have escaped your professional observation that there are only two classes of mankind in the world—doctors and patients. I have had some delicacy in confessing that I have belonged to the patient class ever since a doctor told me that all patients were phenomenal liars where their own symptoms were concerned" [1]. His feelings for physicians prompted the "privilege" (and power) observations above.

Do you think that the physician privilege and power described by Kipling has waned over the last century? I don't think so. At least not much.

For my own part, although I have never hoisted a yellow quarantine flag or ordered a whole neighborhood to be pulled down, I have witnessed bystanders move aside when I happened upon the scene of sudden illness or injury. I recall times when some New York state troopers who knew my car overlooked my driving a few miles above the posted speed, assuming I was, of course, on an emergency call. And when I was in the US Public Health Service serving as a doctor on a Coast Guard cutter, I was put in a small boat and dispatched aboard a huge cargo ship—in rough seas, at night, I might add—which stopped dead still in the Atlantic Ocean while I treated a patient with abdominal pain.

My point to be made, especially to young physicians, is this: The prerogatives described above are yours because of the accomplishments of generations of wise physicians that have preceded you—the giants upon whose shoulders you stand. You have, so far, done little or nothing to justify the lofty position you now hold in society. Thus part of your job description, and one of your long-term career challenges, must be to leave the profession of medicine even more respected, more honorable, than you found it.

This means that, along with striving for excellence in medicine, you must lead lives that inspire confidence in your opinions and recommendations. When you tell a patient that the diagnosis is appendicitis, that surgery is needed, and that you will do the operation, that patient needs to believe in *you*.

What does this mean for new physicians? It means the end of youthful escapades; no more acting like college girls and boys. You are a physician. People trust you with their health and sometimes their lives. Be the physician your patient needs you to be. Act in ways that justify the privilege and power you are accorded, just as Kipling would have you do.

1. Kipling R. A doctor's work. In: A book of words. New York: Doubleday, 1928; pages 44–45.

The Essential Unit of Medical Practice

> The essential unit of medical practice is the occasion when, in the intimacy of the consulting room or sick room, a person who is ill, or believes himself to be ill, seeks the advice of a doctor he trusts.
>
> British pediatrician Sir James Calvert Spence (1892–1954) [1]

By quoting Sir James Spence, we continue with the topic of trust in the healer. Trust is the glue that cements the patient-physician relationship; it is what makes the best doctoring possible. In general, patients trust their physicians, even on first visits. But this general statement must not be swallowed whole, because the bond of trust between patient and physician is being sorely tested by events of our time.

My patient was an elderly man in the hospital with severe heart failure. It was Christmas day and 6 in. of new snow lay on the ground. But dutifully, after our children had opened their gifts, I warmed up the car and drove the 18 miles to the hospital to see my patient. Upon returning home, I called his daughter to report progress. Her first words were: "Thank you, Doctor. I knew you would get to the hospital to see Dad."

That was about 40 years ago, in my solo rural practice days. The fee I received for the hospital visit that day would never compensate for the time away from my family on the holiday, nor for the risk of driving on snow-packed roads. But the expression of trust by the patient's worried daughter was priceless.

Mechanic has observed, "Trust is encouraged by patient choice, continuity of care, and encounter time that allows opportunities for feedback, patient instruction, and patient participation in decisions" [2]. We physicians still enjoy considerable trust. A study of 292 patients revealed that 69 % trusted their personal physicians "to put their needs above all other considerations" [3].

But consider the words of medical educator David E. Rogers (1926–1994): "Individual patient trust, confidence, and comfort have fallen to an alarming low in the UK and the USA. The schizophrenia noted in the polls a number of years ago (i.e. that patients generally have considerable confidence in their own doctor, but a singular distrust of the profession as a whole) continues. However, even an individual patient's confidence in his or her own personal physician is now often more conditional and tenuous" [4].

Why is trust being eroded? Here are just a few reasons: the intrusion of the insurance industry into health care and the suspicion that physician decisions may be driven by profit motives; the publicity about pharmaceutical gifts to physicians and how they may influence prescribing; the breakdown of continuity in the process of health care; and rushed, harried physicians.

A 2013 survey of 1.5 million persons in England revealed an important insight into what fosters trust in the personal physician. The researchers found that among factors considered, "being taken seriously has the strongest association with trust and confidence" in general practitioners [5].

There are many positive examples that enrich the medical profession's treasury of respect and trust—health care teams that serve in disaster areas and impoverished countries, physicians who practice in volunteer clinics helping the homeless, and the countless instances when doctors make extra efforts to stay in contact with patients and offer support during difficult times. Those who do these things may not write about them, but they are the "practice giants" of our day.

1. Spence JC. The purpose and practice of medicine. London: Oxford University Press; 1960,273.
2. Mechanic D. Changing medical organization and the erosion of trust. Milbank Q. 1996;74:171.
3. Kao AC et al. Patients' trust in their physician: effects of choice, continuity, and payment method. J Gen Intern Med. 1998;13:681.
4. Rogers DE. On trust: a basic building block for healing doctor-patient relationships. J R Soc Med. 1994;87(suppl 22):2.
5. Croker JE et al. Factors affecting patients' trust and confidence in GPs: evidence from the National GP patient survey. BMJ Open. 2013;3:e002762.

Medicine as a Social Science

The task of medicine is to promote health, to prevent disease, to treat the sick when prevention has broken down and to rehabilitate the people after they have been cured. These are highly social functions and we must look at medicine as basically a social science.

Swiss-American medical historian Henry E. Sigerist (1891–1957) [1]

Is medicine truly a social science? The term "social science," used as early as the mid-nineteenth century by French philosopher August Comte (1798–1857), is currently considered a broad and ill-defined category that includes anthropology, economics, psychology, sociology, and more—depending on whom you ask. However, today the boundaries of social science, admittedly expansive, are not generally considered to include medicine. Medicine, considered to be the applications of scientific principles together with the art of clinical practice, classically involving one physician and one patient, seems not to fit into the twenty-first century notion of the social sciences.

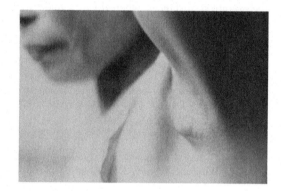

Fig. 1.5 Axillary bubo of bubonic plague. http://www.springerimages.com/Images/MedicineAndPublicHealth/2-AID02E3-05-001

On the other hand, consider the impact of disease on society. The plague of Athens in the fifth century BCE, the cause of which is still debated, killed a quarter of the population. The bubonic plague, the "black death," wiped out a third of the known world's population in the fourteenth century (see Fig. 1.5). The conquest of America was enabled, in many instances, by the deaths caused by newly introduced diseases such as measles in infection-naïve native populations. The 1793 yellow fever outbreak in Philadelphia killed one in ten persons, and influenced the decision to move the US capital out of that city [2]. The 1918 influenza epidemic, one of the most lethal pandemics in history, cased more than 20 million deaths worldwide and greatly augmented the suffering caused by World War I.

And these are just some of the great plagues. Also consider the societal impact of coronary heart disease, stroke, chronic lung disease, arthritis, Alzheimer disease, and more. Disease has also influenced some of the great decisions of history; consider the 1945 Yalta conference, to discuss Europe's fate following World War II, in which the USA was represented by a seriously impaired Franklin Delano Roosevelt, who died three short months later.

Thus, while medicine may not fit a dictionary definition of "social science"—the study of human society and social relationships—there is no denying the impact of various diseases on society and the influences of those who worked to relieve those suffering their manifestations.

1. Sigerist HE. Civilization and disease. New York: Phoenix Books;1962.
2. Rush B. Observations on the origins of the malignant bilious, or yellow fever, in Philadelphia. Philadelphia: Thomas Dobson;1799.

From Aesculapius to Evidence-Based Medicine

Asklepios, god of healing, had two daughters, Panacea (all heal) goddess of healing or clinical medicine, and Hygeia (health or hygiene) goddess of good health, public health, or preventive medicine. The god, like many fathers, hoped that his children would work together harmoniously, but they frequently competed more than they cooperated. If Hygeia were ever to be wholly successful in promoting health and preventing illness, what would there be for Panacea to do? Yet if Panacea dominated medical and popular thinking, who would listen to Hygeia's valuable but often trite strictures? "Eat less, drink less, exercise prudently, or fall into the hands of my sister Panacea and her physicians."

British psychiatrist Humphry Osmond (1917–2004) [1]

In the essay cited above, Osmond (who, parenthetically did pioneering work on hallucinogenic drugs and coined the word *psychedelic*) reminds us that in

Fig. 1.6 Aesculapius, the Greek god of healing (Public domain). http://commons.wikimedia. org/wiki/File:Roman_-_ Statue_of_Aesculapius_-_ Walters_2375.jpg

ancient Greece, there was a third school of medical thinkers, the Dogmatists, who lacked godly parentage. Dogmatists held that medicine must derive from known, articulated principles, and that the Panacean clinicians were merely empiricists [1] (see Fig. 1.6).

Today as the three schools of medicine aspire to act as one, there seems less tension than in the past. We treat malaria, tuberculosis, and acquired immunodeficiency syndrome (AIDS) while working diligently to perfect useful vaccines. Yet, today's darlings seem to be what the Greeks called the Dogmatists, the disciples of evidence-based medicine (EBM). Who can argue against making clinical decisions based on the best current evidence?

But before we fully disregard the value of empiricism and intuition in healers and disease preventers, let us consider the merits of some of the evidence presented to us today. Abraham and Starfield describe the recent dramatic rise in industry-funded studies and their subsequent publication in the medical literature. They tell that approximately 75 % of clinical trials published in *The Lancet*, the *New England Journal of Medicine*, and the *Journal of the American Medical Association* (JAMA) are industry-funded [2]. I find this both astounding and disturbing. They cite the study, published in JAMA, telling that celecoxib (Celebrex), when used for 6 months, is associated with a lower incidence of complications compared with other nonsteroidal anti-inflammatory drugs (NSAIDs), only to find later that unreported data belied the published conclusion [3].

Abraham and Starfield advise, "So what are dedicated clinicians to do? The first step is to give up the illusion that the primary purpose of modern medical research is to improve Americans' health most effectively and efficiently. In our opinion, the primary purpose of commercially funded clinical research is to maximize financial return on investment, not health" [2].

Thus, dear reader, whether you are a follower of the school of Panacea, Hygeia, or the Dogmatists, do not become so cynical that you jettison all confidence in published clinical research. But do pay careful attention to "author disclosures" and the small print telling who paid for studies that might change what you do in practice.

1. Osmond H. God and the doctor. N Engl J Med. 1980;302:555.
2. Abramson J, Starfield B. The effect of conflict of interest on biomedical research and clinical practice guidelines: Can we trust the evidence in evidence-based medicine? J Am Bd Fam Med. 2005;18;414.
3. Silverstein FE et al. Gastrointestinal toxicity with celecoxib vs. nonsteroidal anti-inflammatory drugs for osteoarthritis and rheumatoid arthritis: The CLASS study: a randomized controlled trial. JAMA 2000;284:1247.

Power and the Dream of Reason

> The dream of reason did not take power into account. The dream was that reason, in the form of the arts and sciences, would liberate humanity from scarcity and the caprices of nature, ignorance and superstition, tyranny, and not the least of all, the diseases of the body and the spirit.... Modern medicine is one of those extraordinary works of reason: an elaborate system of specialized knowledge, technical procedures, and rules of behavior.
>
> American sociologist and author Paul Starr (1949–) [1]

Starr's Pulitzer-prize-winning book, *The Social Transformation of American Medicine,* was published in 1982; it tells the tale of medicine in America, its transformation into an industry, and the roles of corporations and government at the time the book was written. The memorable first sentence gives us a hint of things to come in the book and later—about how power shapes events.

What about medicine and power? British philosopher Bertrand Russell (1872–1970) had this to say about power: "The fundamental concept in social science is Power, in the same sense in which Energy is the fundamental concept in physics" [2]. In the struggle for control of America's health care, a tempting prize that accounts for 17 % of the country's gross national product, the power players are physicians, other health care providers, hospitals, pharmaceutical companies, health insurance companies, and government. And, oh yes, patients.

Today, three decades after Starr's book was written: Physicians are struggling to maintain their professional status, often as hourly wage employees of health care systems; other health care providers, nurse practitioners and physician assistants, are competing for increased autonomy and fees; hospitals are expanding their domains and buying clinical practices; pharmaceutical companies continue to use all legal, if not always ethical, methods to increase profits; health insurance companies are raising fees to its customers and struggling to implement a host of new regulations; and patients have continued to be voiceless pawns in all of this. And what of government?

The final page in Starr's book tells us: "Instead of public financing for prepaid plans that might be managed by subscribers' chosen representatives, there will be corporate financing for private plans controlled by conglomerates whose interests will be determined by the rates of return on investments. That is the future toward which American medicine now seems to be headed." At least that was the apparent direction of things in 1982.

But I think that Starr did not take the power of government into account. In 2010 the Affordable Care Act (ACA) became law, the bill regarding which Speaker of the House of Representatives Nancy Pelosi famously said: "We have to pass the bill so that you can find out what is in it, away from the fog of controversy" [3]. We are now, item-by-item, finding out what it contains: Whether you like the law or not, it represents the greatest transformation of the profession and exercise of raw power in the history of American medicine.

1. Starr P. The social transformation of American medicine. New York: Basic Books; 1982.
2. Russell B. Power: a new social analysis. New York: Norton; 1938.
3. Pelosi N. Quoted in: Washington Post. Available at: http://www.washingtonpost.com/blogs/post-partisan/post/pelosi-defends-her-infamous-health-care-remark/2012/06/20/gJQA-qch6qV_blog.html.

Professionalism and Compassion

> I knew a doctor once who was honest, but gentle with his honesty, and was loving, but careful with his love; who was disciplined without being rigid, and right without the stain of arrogance; who was self-questioning without self-doubt, introspective and reflective and in the same moment, decisive; who was strong, hard, adamant, but all these things laced with tenderness and understanding; a doctor who worshipped his calling without worshipping himself; who was busy beyond belief, but who had time—time to smile, to chat, to touch the shoulder and take the hand; and who had time enough for Death as well as Life.
>
> American physician and writer Michael A. LaCombe, MD (1942–) [1]

At the beginning of this chapter, I discussed the concept of professionalism, including the current "charter," "blueprint," and guidelines for assessing competence.

As I read the words above by LaCombe, I wonder if his is the true description of professionalism—describing all the traits that we, as healers, wish we had. If I had to sum up the attributes LaCombe ascribes to his model physician in one word, it would be *compassion*. Crawshaw describes it this way, in a metaphoric saying he attributes to an "elderly Russian physician": "But of course, without *sostradanya* (compassion) no man is a doctor. A doctor must give part of his heart to his patient" [2].

At this point, I reviewed the document *Medical Professionalism in the New Millennium: A Physician Charter*, searching the ten commitments; here I found commitments to honesty, patient confidentiality, scientific knowledge, and maintaining trust by managing conflicts of interest [3]. There is no specific mention of compassion, of the qualities described by LaCombe. Nor is compassion found in the categories measured in the blueprint to assess professionalism [4].

Why the absence of compassion in these documents? Is the omission a conscious acknowledgement that the compassion LaCombe describes is an outlier, representing an ideal few physicians can achieve? Or do we assume that compassion is something all physicians possess, and that its inclusion in formal descriptions of professionalism would somehow be redundant?

There is hope for the future. There is a new tool some medical schools are using to evaluate applicants to medical schools—the multiple mini-interview (MMI) process, which uses a series of "stations" instead of traditional interviews. One of the personal traits assessed in the MMI is compassion [5].

1. LaCombe MA. On professionalism. Am J Med. 1993;94(3):329.
2. Crawshaw R. Humanitarianism in medicine. In: Smith MEC. Living with medicine: a family guide. Washington DC: Am Psychiatry Assoc Aux;1987.
3. Medical professionalism in the new millennium: a physician charter. Ann Intern Med. 2002;136:242.
4. Wilkinson TJ et al. A blueprint to assess professionalism: results of a systematic review. Acad Med. 2009;84:551.
5. Kirsch DG. Transforming admissions: the gateway to medicine. JAMA. 2012;308:2250.

Professionalism, Quality, and Social Benefits

> Perhaps most important, professionals have an ideology that assigns a higher priority to doing useful and needed work than to economic rewards, an ideology that focuses more on the quality and social benefits of work than its profitability. Although this is the most important part of medical professionalism, it is now what is most at risk.
>
> American physician and journal editor Arnold S. Relman, MD (1923–2014) [1]

Arnold Relman was editor of the *New England Journal of Medicine* from 1977 to 1991, and he has published prolifically on topics such as medical writing and health care in America. In the article cited above, published in 2007, Relman goes on to comment: "A major reason for the decline of medical professional values is the growing commercialization of the US health care system" [1].

When I opened my solo practice office in a more-or-less rural county of upstate New York in 1968, I checked with the county medical society about the propriety of placing a discrete announcement in the town newspaper. They approved, within certain guidelines.

Then in 1972, I published my first book, a health care guide for the senior citizen titled *Feeling Alive after 65*. When the local radio station invited me to come for an interview, the county medical society needed to discuss whether or not this would be a breach of professionalism. They eventually decided that my discussion of health issues on the radio was educational and not advertising, and I made my first media appearance.

Today I checked the local *Yellowbook* of telephone numbers. Under "Physicians— MD & DO" I found display ads for physicians with "Excellence in the care of women with urinary control, overactive bladder, and pelvic problems," and another for a physician advertising herself as an "Adult Specialist." Is the being an adult a new disease, or is the physician herself now an adult?

When driving, I see billboards inviting me to have chemical peels and Botox injections administered by doctors. My radio advises me of various physicians' special skills in joint replacement or in the care of heart disease, supported by patients testifying to the outstanding treatment received. A North Carolina hospital recently received some criticism for its new slogan *"Cheat Death"* [2].

Osler said: "The practice of medicine is an art, not a trade; a calling, not a business." In this modern world of competition, conspicuous consumption, integrated health care systems, and health insurance exchanges, we must resist the temptation to become very well compensated, but humbly obedient servants of the evolving health care hegemony. If we can just hold fast to our service-based ideology and our professional values, we will still continue to enjoy very good incomes and lives, and perhaps we can escape the often-corrupting influence of today's medical marketplace. That is, we will continue to behave as professionals, and not as used car salesmen.

1. Relman AS. Medical professionalism in a commercialized health care market. JAMA. 2007;298:2668.
2. DePriest J. "Cheat Death" slogan draws scrutiny to Gaston hospital. The Charlotte Observer: Local news. Posted April 5, 2013. Available at: http://www.charlotteobserver.com/2013/04/05/3960526/cheat-death-slogan-draws-scrutiny.html.

The Physician's Duty and Continuity of Care

We physicians are like the sea at city's edge—a point of orientation. We are a beacon for patients in crisis. We are a locker for their sunken histories. We gather flotsam from their foundered dreams. Knowing this is to know something of our duty, something about the continuity of care.

America family physician David A. Loxterkamp, MD (1953–) [1]

Here is my story about continuity of care; only the names are changed. I entered private practice in 1964 as the freshman member of a four-doctor family practice group in a small town in upstate New York. During my first week in this practice, it was my afternoon to do the group's house calls. The first visit was to an irascible 75-year-old Italian man, Mr. Fontana, patriarch of his family, sick in bed with a high fever. As I entered his bedroom, he challenged me: "Who are you? My doctor is Dr. Venanzi. Where is he? I don't want to see anyone but him."

"Sorry," I replied, "Dr. Venanzi has appointments in the office all day, and I am assigned to house calls." But the patient was adamant, and so I turned to leave. But as I opened the door, he called from the bedroom, "Wait a minute. You'll do!"

Somehow we hit it off. Subsequently, Mr. Fontana became my patient, as did his wife, children, and grandchildren, and they remained my patients for years. He, his family, and I developed a sense of continuity, what Balint called the "mutual investment company" [2]. In fact, my experience with Mr. Fontana and his family showed me the two faces of continuity. It taught me clinical benefits of longitudinal care and yet just how fragile trust can be when the patient perceives that the continuity contract has been breached, as Mr. Fontana concluded when Dr. Venanzi failed to come when he was sick.

Continuity of care is challenged today by urgent care centers, part-time practitioners, hospitalists, nocturnalists, and laborists (physicians who do only deliveries on shift work). Often the two-income family with small children must seek medical care at off hours and wherever care is available when required. Despite what patients like Mr. Fontana might wish, continuity of care is often trumped by convenience—of both patient and doctor.

There are many reasons why we should combat this trend. Following a critical review of the literature, Saultz and Lochner conclude that it is "likely that a significant association exists between interpersonal continuity and improved preventive care and reduced hospitalization" [3]. A study of 21,698 patients at Veterans Affairs medical centers demonstrated the high value patients place on continuity of care; the authors of the study found "that continuity of health care was strongly associated with higher satisfaction of patients with the humanistic skills of their primary care provider, and with the organization and access to care" [4].

In clinical encounters of all types, knowing the patient, the family, and all the bits and pieces of past history makes it more likely that the correct diagnosis will result, that care will be rendered more expeditiously, and that everyone will be more satisfied than would occur with ad hoc visits to whatever provider is available.

1. Loxterkamp DA. Old men and the sea. JAMA. 2010;304:18.
2. Balint M. The doctor, his patient, and the illness. New York: International Univ. Press; 1072, page 133.
3. Saultz JW et al. Interpersonal continuity of care and care outcomes: a critical review. Fam Med. 2005;3:159.
4. Fan VS et al. continuity of care and other determinants of patient satisfaction with primary care. J Gen Int Med. 2005;20:226.

Chapter 2
Being a Physician

Imagine that you were interviewing applicants for medical school—which I happen to do—and you were looking for the ideal future physician. What would you consider the most desired attributes? From the standpoint of a medical educator, I would want the applicant to be bright and inquisitive, an eager learner who will participate actively in class discussions and who will bring new ideas to the group. As a practicing physician, I want my future colleague to be reliable, well educated and trained, morally and ethically grounded, and imbued with the team spirit needed to make medical practice—whether group or solo—successful and satisfying for patients, staff, and colleagues.

What if you were the patient? As a sometime patient, I want my physician to be clinically competent, to be compulsive when it comes to following up on reports, to

© Springer Science+Business Media New York 2015
R.B. Taylor, *On the Shoulders of Medicine's Giants*,
DOI 10.1007/978-1-4939-1335-0_2

Fig. 2.1 The stethoscope: one of the enduring symbols of medicine and the physician. This image found at: http://www. springerimages.com/Images/ MedicineAndPublicHealth/ 1-10.1007_978-1-84882-515-4_21-0

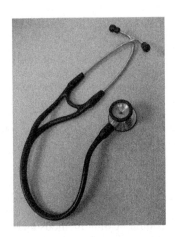

be innovative in searching out new ways to approach old problems, and, perhaps most of all and even if it seems a little old-fashioned, to care for me as a person.

None of the attributes described above are mutually exclusive, and all are consistent with being an outstanding learner, colleague, and healer. Together, they describe qualities important in being a physician (see Fig. 2.1).

In this chapter, I will discuss the physician as a healer and as a person. Topics covered include the health of the physician as well as the health of the patient; service and giving back to the world; what makes a "complete" physician; and when to hang up the stethoscope and declare one's career is over.

Protecting the Patient

The patient, who may mistrust his own parents, sons and relations, should repose an implicit faith in his own physician, and put his own life into his hands without the least apprehension of danger; hence a physician should protect his patient as his own begotten child.

Suśhruta (ca. sixth century BCE) [1]

An Indian sage Suśhruta, who probably lived a century or so before Hippocrates, helped develop the *Ayurveda* system of lifestyle and medicine. His works included—remarkably—descriptions of several hundred surgical techniques such as dilation of urethral strictures, hernia repair, and birth by caesarean section.

I find several intriguing aspects to the quote above. First, although Suśhruta and Hippocrates lived and practiced in the same general era, Hippocrates advocated natural healing methods and was not a strong proponent of operative intervention. Suśhruta, on the other hand, described surgery, and is quoted as telling, "Surgery is the first and the highest division of the healing art, pure in itself, perpetual in its applicability, a working product of heaven and sure of fame on earth" [2].

Second, it is perhaps no coincidence that, as a surgical "giant," Suśhruta's quotation above stresses protection of the patient. Keeping the patient from harm is just as pertinent now as it was in ancient India. Here are just some of the dangers:

- Errors of diagnosis: A pigmented skin lesion considered to be a simple mole turns out to be a melanoma. An infant with a fever attributed to a simple viral infection is found in the morning to have a stiff neck and ecchymotic rash. Errors of diagnosis are distressingly common and, according to Ely et al., can often be linked to the cognitive biases and perilous mental shortcuts of clinicians, who, in many cases, fail to consider the correct cause in the differential diagnosis [3].

- Inappropriate surgery: This is not about amputating the wrong leg, but about surgery that is not really needed. In the USA there are about 1 million hip and knee replacements annually, which is almost one in every 300 Americans. Yet there are no specific guidelines as to when major joint arthroplasty should be performed [4].
- Ridiculous recommendations by unqualified persons: A physician recently reported that a patient told of being advised by her yoga instructor to drink her own urine each morning because it would "cure everything" [5]. Today we read a lot about complementary and alternative medicine (CAM), and some of our best young medical practitioners have become passionate advocates, despite its mystical roots and paucity of rigorously vetted evidence.
- Abuse by insurance providers: This includes both private and government sponsored programs. What happens is a little like home insurance; you sign up optimistically expecting fair treatment and then when disaster occurs, you learn about the exclusions. Fighting insurance providers is tedious and time-consuming, which is what the other side wants, but it is the physician's duty today to be the spokesperson for patients who don't know the pathways to the goal of fair treatment.

I could name more dangers patients face, such as pharmaceutical errors and rushed hospital discharges. It seems to me that, more than ever, today's patients need physicians to offer protection as part of their care.

1. Sharma PV. Suśhruta-Samhitā (Varanasi: Caukhambha Visvabharati). 2000; vol. 1.
2. Rajan S. Who is the father of surgery in India? Available at: http://www.sukh-dukh.com/forums/showthread.php?t=79736.
3. Ely JW et al. Checklists to reduce diagnostic errors. Acad Med. 2011;86:307.
4. Lawson EH et al. Appropriateness criteria to assess variations in surgical procedure use in the United States. Arch Surg. 2001;146:1433.

Courage in Times of Crisis

At the beginning (of the plague of Athens in the fifth century BCE) the doctors were quite incapable of treating the disease because of their ignorance of the right methods. In fact, mortality among the doctors was the highest of all, since they came more frequently in contact with the sick.

Greek general and historian Thucydides (c. 460–395 BCE) [1]

One of the quiet little secrets of medicine is that doctors and others who provide direct patient care are exposed to a lot of very sick persons, some with infections that defy antibiotics and some patients with mental derangements that make them dangerous to themselves and others, including those who provide health care.

In 1883 Dr. F.R. Hudson was shot and seriously wounded by a man whose wife he declined to visit [2]. In 2013 in Beijing, a doctor was stabbed to death by a patient displeased with the outcome of surgery on his nose [3]. Attacks on physicians are

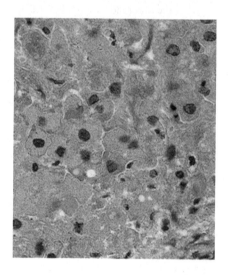

Fig. 2.2 Ebola virus: Hepatocellular necrosis and typical intracytoplasmic viral inclusions are seen in liver of a patient with Ebola virus hemorrhagic fever. http:// www.springerimages.com/ Images/Medicine AndPublicHealth/2- AID08E3-10-048A

not uncommon and are among the hazards that attend the practice of medicine. But more to the point are the infectious diseases that we can treat but cannot cure.

O'Flaherty, in 1991, wrote of the physician's duty to treat the patient with communicable disease. In the setting of some physicians refusing to treat AIDS patients, she wrote: "Physicians today, inexperienced at weighing personal risk against professional responsibility, are examining the extent of their occupational obligation" [4].

Over the past two decades, we have become more comfortable treating patients with AIDS. We better understand its modes of transmission, we have refined our methods of avoiding contagion, and the drugs are superior. But what of future plagues?

Consider the Ebola and Marburg viruses, and other causes of viral hemorrhagic fever for which there is no immunity, no reliable prophylaxis, and no sure cure (see Fig. 2.2). When these highly contagious infections hitchhike to our world aboard a sick airline passenger or are introduced by a terrorist, how many among us will follow the example of Benjamin Rush M.D. who, during the 1793 yellow fever epidemic in Philadelphia, wrote to his wife: "I had resolved to perish with my fellow citizens rather than dishonor my profession or religion by abandoning the city" [5].

Who would enter the medical profession if the risk to the practitioner's life were high? Happily, this is not the case today. We continue to find innovative ways to protect physicians, staff, and their families from infectious diseases, and, unless some catastrophic, communicable plague emerges, health care professionals are in less danger from their patients than from lifestyle choices we make for ourselves each day, as discussed in the next section.

1. Thucydides: History of the Peloponnesian War, Warner R, trans. Middlesex, England: Penguin Books; 1954,151.
2. News Note: The Journal;1883;1:281.
3. Chinese doctor stabbed to death in latest hospital attack. Available at: http://news.yahoo.com/chinese-doctor-stabbed-death-latest-hospital-attack-084856196.html.
4. O'Flaherty JO. The AIDS patient: a historical perspective on the physician's obligation to treat. The Pharos. 1991;54;13.
5. Butterfield LH, ed. Letters of Benjamin Rush, vol. 2: 1793–1813. Princeton: Princeton University Press;1951,734.

About the Health of the Physician

The physician will hardly be thought very careful of the health of others who neglects his own.

Roman physician and philosopher Claudius Galen (ca. 130–200) [1]

He was known as Galen of Permegon, a city in what is now Turkey, where he was born. He was a leading physician of ancient Rome, with a special interest in anatomy, perhaps based on his experience treating the wounds of gladiators and his dissection of animals. He systematized the medical knowledge of his day in hundreds of written works, and his influence lasted for 1,500 years (see Fig. 2.3).

However, Galen was wrong on several important points. As examples, he believed that arterial blood and venous blood were contained in separate systems, and that humans have two-chamber hearts and five-lobe livers. His work has been described, perhaps not totally fairly, as "a weird hodgepodge of nonsense, Aristotelian philosophy, Hippocratic dogma, and shrewd clinical and experimental observations" [2].

Fig. 2.3 Galen of Permegon.
http://www.springerimages.
com/Images/Medicine
AndPublicHealth/
5-10.1186_1748-7161-4-6-10

But Galen was correct when it came to physicians preserving their own health. The admonition goes beyond avoiding contagious diseases. Today the greater threats to physicians are depression, substance abuse, premature diseases of aging, and suicide.

Gerber writes: "A troubled physician and physician's family is one result of working hard for others and ignoring one's own needs" [3]. The overburdened physician, often depressed, may be tempted to self-treat with alcohol or drugs.

Gastfriend reports a 10–15 % prevalence of substance abuse disorders among physicians; that this prevalence mirrors that in the general population should give us scant comfort [4].

With practice pressures, long work hours, and often sleep deprivation not unlike the experience of residents in training, physician well-being can yield to career exhaustion, and studies show that, at any time, approximately one in three physicians is experiencing burnout [5].

To combat these problems, the Mayo Clinic has pioneered the "Physician Well-Being Program" to promote physician well-being through research, education, and the development of wellness promotion programs that foster physician satisfaction and performance" [6]. But substantive change can only follow a fundamental philosophical change in physician thinking.

We clinicians must come to believe that only when we are emotionally and physically healthy—and rested—can we provide optimum care. Anything less, neglecting our own health, is in fact being careless with the health of others.

1. Claudii Galeni Pergameni. Of protecting health, book 5.
2. Inglis B. A history of medicine. New York: World Publishers; 1965, page 39.
3. Gerber LA. Married to their careers: career and family dilemma in doctors' lives. New York: Tavistock; 1983, page 81.
4. Gastfriend DR. Physician substance abuse and recovery: what does it mean for physicians—and everyone else? JAMA. 2005;293:1513.
5. Shanafelt TD. Enhancing meaning in work: a prescription for preventing physician burnout and promoting patient-centered care. JAMA. 2009;302:1338.
6. Physician Well-being Program, Mayo Clinic: Available at: http://mayoresearch.mayo.edu/mayo/research/physicianwellbeing/.

Ceasing Doctoring at the Right Time

I have entered my eighty-fifth year; and when I retired a few years ago from the practice of physic, I trust it was not a wish to be idle, which no man capable of being usefully employed has a right to be; but because I was willing to give over before my presence of thought, judgment and recollection was so impaired that I could not do justice to my patients. It is more desirable for a man to do this a little too soon, than a little too late; for the chief danger is on the side of not doing it soon enough.

English physician William Heberden (1710–1801) [1]

For many reasons, physicians are retiring earlier today, according to hospital executives, physician recruiters, and researchers in the field [2]. Among the reasons cited by physicians, the commercialization of medicine tends to lead the list. But when should you and I hang up our stethoscopes and actually retire?

Fig. 2.4 William Heberden. http://www.springerimages. com/Images/MedicineAnd PublicHealth/1-10.1007_978- 1-84882-342-6_1-2

According to Heberden, one of leading physician scholars of his day, for whom the knobby distal interphalangeal joint manifestations of arthritis are named (see Chap. 5), a wise clinician retires before becoming a danger to patients (see Fig. 2.4). In sports parlance this might be considered retiring at the top of your game.

In my early years of practice I encountered a beloved, quite elderly physician who was still seeing his long-time patients, even though his clinical skills and judgment had retreated to a danger zone. The physician leader at the local hospital, upon being advised of the problem by colleagues, had a serious chat with his aged colleague, who soon afterwards began his well-deserved retirement.

One valuable option in physician retirement today is the "bridge job" [3]. The previously held belief that the practice of medicine is "all or nothing" need not be true today. Whatever one's practice style—entrepreneur, employee, or academic—the ability to identify medically related activities that allow phased retirement can be the key to contentment once full-speed practice is in the rear-view mirror. Bridge job possibilities include administration, insurance medicine, or volunteer care of the homeless.

As an academic physician, I "stepped down" as department chairman at age 62, and I continued to see patients for the next 8 years and spent my nonclinical time teaching, writing, and mentoring young faculty. Like any responsible faculty dinosaur, I progressively reduced my weekly work hours, until I finally "retired" at age 76; but I still volunteer as an interviewer for applicants to the local medical school and I write books like this.

I believe that I did not retire to be idle, and that my "bridge jobs" were and are useful to society. Yet I ceased active patient care before, in Heberden's words, "my presence of thought, judgment and recollection was so impaired that I could not do justice to my patients" and before a major medical misadventure occurred.

1. Heberden W. Letter. Quoted in: Scarlet EP. Fair flower of Harvard. Arch Int Med. 1965;116:611.
2. Bahrami B et al. Is there any difference in the retirement intentions of female and male physicians? Clute Institute J Bus Econ Res. 2004;2:45.
3. Kim S et al. Working in retirement: the antecedents of bridge employment and its consequences for quality of life in retirement. Acad Management J. 2000;43:1195.

The Self-Complacency of the Young Physician

> The first acts of a graduate are apt to be his precedents through the coming years, for there is no era in life in which his self-complacency is so exalted as the time which passes between receiving his diploma with its blue ribbon, and receiving cape and gloves to wear at the funeral of his first patient.
>
> American physician and educator Daniel Drake (1785–1852) [1]

Daniel Drake was a towering figure among American physicians. He received the first medical diploma in Cincinnati in 1805, the first awarded west of the Appalachian Mountains. Drake was instrumental in establishing the Medical College of Ohio and he served as its first president and, later, led the Ohio State Medical Society. He also, in 1827, founded the *Western Journal of the Medical and Physical Sciences*, which he continued to edit until 1848 [2] (see Fig. 2.5).

Fig. 2.5 Daniel Drake (Public domain). http://commons. wikimedia.org/wiki/ File:Daniel_Drake002.png

Could it be that the "self-complacency," perhaps the self-confidence, described by Drake characterizes the frontier medicine of his day. Courage was often needed when resources were in short supply. Remember that we would not see ether anesthesia until 1846, Lister's antisepsis would come two decades later, and the first targeted antimicrobial, Ehrlich's arsenic-based anti-syphilitic drug Salvarsan, would eventually follow in 1910.

As a young physician practicing in Ohio in the early nineteenth century, Drake seems to have been a risk-taker, living by the Thomas Aquinas adage; "If the highest aim of a captain would be to preserve his ship, he would keep it in port forever." Aquinas would probably applaud the physician who takes reasonable risks in the best interests of his patients—the heroic surgery or the trial of an untested drug when all else has failed.

My colleague, Bob Bomengen, M.D., a solo physician practicing in an Oregon frontier community, was faced with a 14-year-old boy accidentally shot in the abdomen, with the aorta damaged. "Dr. Bob" opened the abdomen and, with his hand, prevented aortic hemorrhage until the rescue helicopter could arrive from 175 miles away. The boy survived [3]. Another family physician colleague, early in his practice, suddenly faced with a patient with acute subarachnoid hemorrhage, used the instruments available in his office to drill a hole in the patient's skull; the patient lived, presumably owing to the emergency surgery. What courage—self-confidence— these acts must have taken?

And then comes the first preventable death, a sobering event for a young doctor doing his best for his or her patients. Most physicians can recall this first patient death decades later. Today, with postgraduate training programs of 3 to even 8 years, most young physicians experience their first patient death during training. The wise physician learns from the experience and goes on, but a little more cautiously than before.

1. Drake D. Editorial. West J Med Surg. 1844;354:2.
2. Daniel Drake: In: Appleton's cyclopedia of American biography. New York: Appleton; 1887, p. 223.
3. Reader's Digest. May, 2005.

The Physician as Advocate

Only those who regard healing as the ultimate goal of their efforts can, therefore, be designated as physicians.

German pathologist Rudolph Virchow (1821–1902) [1]

Since early times, the holy grail of early "medicine men" (and women), and later, physicians, of diagnosis and therapy has been healing. Sometimes it is actual cure; at other times, support while nature heals; and all too often, the healing power of caring when the end of the therapeutic trail has been reached. And at times, part of healing is speaking out for the patient.

Fig. 2.6 Rudolph Virchow. http://www.springerimages. com/Images/Medicine AndPublicHealth/1-10.1007_ s00268-004-2056-0-0

Virchow was a leading scientist of his day (see Fig. 2.6). But his legitimacy to speak of healing comes from his promotion of medicine's role in social reform: "Medicine, as a social science, as the science of human beings, has the obligation to point out problems and to attempt their theoretical solution: the politician, the practical anthropologist, must find the means for their actual solution... The physicians are the natural attorneys of the poor, and social problems fall to a large extent within their jurisdiction" [2].

To Virchow's thinking, part of doctoring is advocacy, helping the patient and family cope with the problems we all recognize and, today, the labyrinthine health care bureaucracy we all face. This is our work: to diagnose, to treat as best we can, and to advocate for our patient when needed. These three phases are all part of healing. The latter—the advocacy championed by Virchow—may seem the least gratifying of the three activities, but sometimes can be the most effective.

Today, there is a growing sense that physician advocacy is fundamental to medical professionalism. Yet, as Earnest et al. state: "Despite widespread acceptance of advocacy as a professional obligation, the concept remains problematic with the profession of medicine because it remains undefined in content, scope, and practice" [3]. One impediment is that patient advocacy activities, such as writing letters to insurance companies when vital medication or surgery recommendations are denied, are unlikely to be "reimbursable" events. When the health care system eventually defines the physician's role to include addressing social problems that affect health and compensates fairly for this type of activity, the lives of both patients and physicians will be better. And advocacy will be formally recognized as an important role of healers.

1. Virchow R. Disease, life and man: selected essays by Rudolph Virchow. Trans. by L. J. Rather. Stanford, CA: Stanford University Press; 1958.
2. Brown T et al. Rudolph Carl Virchow. Am J Public Health. 2006;96:2104.
3. Earnest MA, et al. Physician advocacy: what is it and how do we do it? Acad Med. 2010;85:63.

Being a Pioneer

It is not easy to be a pioneer—but oh, it is fascinating! I would not trade one moment, even the worst moment, for all the riches in the world.

American physician Elizabeth Blackwell (1821–1910) [1]

Elizabeth Blackwell, the first woman to receive an M.D. degree in the USA, was certainly a pioneer. Born in England, Blackwell moved with her family to the USA at age 11. Her decision to pursue a medical career, like the decisions of many aspiring physicians today, was influenced by the painful death of a friend who, we think, suffered from uterine cancer [1, 2] (see Fig. 2.7).

After one rejection after another from leading medical schools of the day Blackwell was finally accepted by Geneva Medical College, now State of New York Upstate Medical University in Syracuse, New York. Even here, her application prompted such indecision among the administrators that her fate was put to a vote of the 150 male students. If even one voted "nay," she would be rejected. But, according to legend, the young men, considering the matter to be comical, voted

Fig. 2.7 Elizabeth Blackwell (Public domain). http://commons.wikimedia.org/wiki/Elizabeth_Blackwell#mediaviewer/File:Elizabeth_Blackwell.jpg

unanimously for her acceptance [3]. In 1849, Elizabeth Blackwell received her medical degree.

Blackwell continued her studies in France, and her career included time in England, where she became the first woman on the United Kingdom Medical Register. In 1851, she returned to the USA where her work included clinical practice, lectures, writing, and social activism [4].

Distinguished as her career was, Blackwell's chief contribution to medicine was serving as a trailblazer for generations of young women who would follow her into medical careers. These included her own sister, Emily Blackwell, who was the third American woman to earn an M.D. degree. Other American beneficiaries of her pioneering efforts include such illustrious names as:

- Annie Lowrie Alexander (1864–1929)—first licensed female physician in a southern US state [5].
- Florence Rena Sabin (1871–1953)—first woman physician elected to the US National Academy of Sciences, discussed later in Chap. 8.
- Marie Equi (1872–1952)—a leader in the fight for women's access to birth control.
- Virginia Apgar (1909–1974)—the "mother" of neonatology and the APGAR score.
- Nancy Dickey (1950–)—first female president of the American Medical Association.

There are thousands more, thanks to the bravery and determination of Elizabeth Blackwell and those who followed her path. Today women constitute almost half of all US medical students.

1. Baker R. The first woman doctor: the story of Elizabeth Blackwell, MD. New York: J Messner; 1944.
2. Roth N. The personalities of two pioneer medical women: Elizabeth Blackwell and Elizabeth Garrett Anderson. Bull NY Acad Med. 1971;47:67.
3. Curtis RH. Great lives: medicine. New York: Atheneum; 1993.
4. Sahli NA. Elizabeth Blackwell, MD (1821–1910). New York: Arno Press: 1982.
5. Cohn S. More than petticoats: remarkable North Carolina women. Guilford CT: Globe Pequot; 2012; p. 82.

Standing Above the Common Herd

There are men and classes of men that stand above the common herd: the soldier, the sailor, and the shepherd not infrequently; the artist rarely; rarer still, the clergyman; the physician almost as a rule. He is the flower (such as it is) of our civilization; and when that stage of man is done with, and only to be marveled at in history, he will be thought to have shared as little as any in the defects of the period, and most notably exhibited the virtues of the race. Generosity he has, such as is possible to those who practice an art, never to those who drive a trade; discretion, tested by a hundred secrets; tact, tried in a thousand embarrassments; and what are more important, Herculean cheerfulness and courage. So that he brings air and cheer into the sick room, and often enough, though not so often as he wishes, brings healing.

Scottish author Robert Louis Stevenson (1850–1894) [1]

Fig. 2.8 Robert Lewis Stevenson. http://www. springerimages.com/search. aspx?caption=Robert%20 Louis%20Stevenson%20

Stevenson, although not a physician, experienced several serious illnesses during his life and thus spent his share of time in the presence of healers. He was both a grateful and articulate patient. The now-famous tribute to physicians quoted above was published, curiously, in the Dedication of a book of children's poems [1]. Being both patient and poet, he offered a special perspective on the characteristics of physicians he had known and their virtues, including a comparison to persons in other life roles. I note with amusement that, when it comes to those who "stand above common herd," Stevenson not unexpectedly places physicians at the top of his list. But along the way, he ranks soldiers, sailors, and shepherds above artists, and all of these above the clergy (see Fig. 2.8).

Despite the skills, professionalism, and altruism of most physicians today, medical doctors no longer seem to "stand above the herd," at least when it comes to public confidence in professional honesty and ethics. In a Gallup Survey using honesty and ethical standards as criteria, medical doctors ranked a modest fifth, trailing nurses, military officers, pharmacists, and grade school teachers. Physicians continue to be ranked above, actually two levels above, the clergy. The bottom of the list? Members of congress, car salespeople, and lobbyists [2].

Interestingly, our US cultural priorities may be revealed in a study from the University of Wisconsin ranking occupational prestige—the relative social class of those in various jobs. In this survey, physicians were listed number one, ahead of attorneys, computer scientists, college professors, and physicists [3]. It makes me wonder if prestige is still equated with honesty and high moral standards, or is it viewed as indicating financial success?

1. Stevenson RL. Dedication. Underwoods: a child's garden of verses. New York: Peter Fenelon Collier & Son: 1887.
2. Gallup Honesty and Ethics Survey: 2010. Available at: http://www.gallup.com/poll/145043/Nurses-Top-Honesty-Ethics-List-11-Year.aspx.
3. Hauser RM, et al. Socioeconomic indexes for occupations: a review, update, and critique. Madison, Wisconsin: Center for Demography and Ecology, University of Wisconsin; 1996.

Hobby-horses, Tyrants, and Mighty Purposes

A man like me cannot live without a hobby-horse, a consuming passion—in Schiller's words a tyrant. I have found my tyrant, and in his service I know no limits. My tyrant is psychology. It has always been my distant, beckoning goal and now since I have hit upon the neuroses, it has become so much the nearer.

Austrian neurologist Sigmund Freud (1856–1939)

We health professionals know about Freud, recognized as the "father of psychoanalysis" (see Fig. 2.9). Less well known to doctors is the man whom Freud cites, German playwright Friedrich Schiller (1759–1805). Freud's allusion to his work, and specifically to the word "tyrant" sent me on a quest for context—the literary work Freud was citing in the quote above. I believe my search was successful. Here is what I found: Describing Schiller's play, *Fiesco: or the Genoese Conspiracy*, Thomas writes: "No one can mistake the autobiographic note in the speech of Bourgognino which closes the first act: 'I have long felt in my breast something that would not be satisfied. Now of a sudden I know what it was. (*Springing up heroically*) I have a tyrant'" [2].

What Freud and Schiller describe as a "hobby-horse" or "tyrant," I think of as a mission, an aim, or a purpose that gets you out of bed in the morning and gives life extra meaning. Here, from the Preface to the drama *Man and Superman* by Irish playwright George Bernard Shaw (1856–1950), is one of my favorite inspirational quotes: "This is the true joy in life, the being used for a purpose recognized by

Fig. 2.9 Sigmund Freud.
http://www.springerimages.
com/Images/MedicineAnd
PublicHealth/1-10.1007_
978-1-4614-0170-4_1-24

Fig. 2.10 Sir Winston Churchill. http://www. springerimages.com/search. aspx?caption=winston churchill

yourself as a mighty one; the being thoroughly worn out before you are thrown on the scrapheap; the being a force of Nature instead of a feverish, selfish, little clod of ailments and grievances complaining that the world will not devote itself to making you happy" [3].

While writing a book like "Medicine's Giants," I sometimes think of another writer, Sir Winston Churchill, who used the word *tyrant*: "Writing a book is an adventure. To begin with it is a toy and an amusement. Then it becomes a mistress, then it becomes a master, then it becomes a tyrant. The last phase is that just as you are about to be reconciled to your servitude, you kill the monster and fling him to the public" [4] (see Fig. 2.10).

How many of us have "hobby-horses?" For most physicians so afflicted, of course, the "consuming passion" is medicine, especially the doctor's specialty. I wonder how many of these "hobby-horses" evolve into "masters," even "tyrants," that dominate the lives of unsuspecting physicians. How many feel the passion, and master it? Or, is this sort of passion seldom seen in young physicians today?

1. Freud S. Letter to William Fless, 1895, Quoted in: J History of Behav Sci. 1967;3–4:159.
2. Thomas C. The life and works of Friedrich Schiller. Middlesex, England: Echo Library; 2006, page 57.
3. Shaw GB. Preface: Man and Superman, a drama. 1903.
4. Winston Churchill quotes. Available at: http://www.goodreads.com/quotes/37949-writing-a-book-is-an-adventure-to-begin-with-it.

Service in Medicine

You ask me to give you a motto. Here it is: Service. Let this word accompany you as you seek your way and your duty in the world. May it be recalled to your minds if ever you are tempted to forget it or to set it aside. Never have this word on your lips, but keep it in your hearts. And may it be a confidant that will teach you not only to do good but to do it simply and humbly. It will not always be a comfortable companion but it will always be a faithful one. And it will be able to lead you to happiness no matter what the experiences of your lives are.

Physician and theologian Albert Schweitzer (1875–1965) [1]

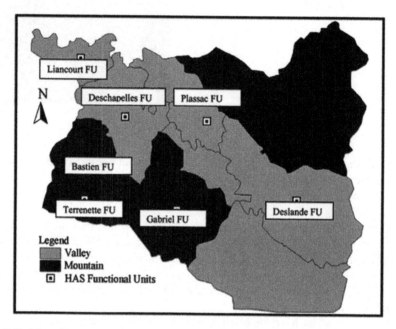

Fig. 2.11 Map of the Functional Units (FU) where the Hôpital Albert Schweitzer (HAS) operates its programs in Haiti. http://www.springerimages.com/Images/MedicineAndPublicHealth/5-10. 1186_1475-9276-6-7-1

Best known for his service as a medical missionary, Schweitzer founded a hospital at Lambaréné, Gabon in 1913. The advice cited was directed to a group of nursing students, but it should resonate with all healers. Certainly the spirit of service was, as described in the previous section, Schweitzer's "mighty purpose."

There is also a Hospital Albert Schweitzer at Deschapelles in Haiti, founded by Larry and Gwen Mellon in 1956 as a tribute to Schweitzer. The hospital is the hub of an integrated health system serving the inhabitants of central Haiti (see Fig. 2.11).

Aside from the laudatory altruism advocated by Schweitzer, there is a moral justification for our obligation to serve others. Despite the mountains of debt many young physicians incur during education and training, none of us got where we are alone. Our parents contributed mightily to our success, as did our teachers through all levels of education. Physicians in many specialties, nurses, technicians, and other health professionals shared their wisdom, their clinical pearls, and their diagnostic maneuvers—all in an effort to make each of us the best healer possible.

The medical school equivalent of undergraduate Phi Beta Kappa is Alpha Omega Alpha (AOA), the professional organization that "recognizes and advocates for excellence in scholarship and the highest ideals in the profession of medicine." Only students in the top 25 % of a medical class can be considered for membership; not all are selected. The AOA motto, since the time of founder William W. Root, M.D. in 1902, is: "Be worthy to serve the suffering" [2]. But one need not be in the top quartile of a medical school class to aspire to this message. Every healer, whatever his or her class rank, should "be worthy to serve the suffering."

The latest initiative to recognize medical service to humanity is the Gold Humanism Honor Society (GHHS), established in 1988 by the Arnold P. Gold Foundation to foster humanism and professionalism in medicine. The GHHS "honors medical students, residents, role-model physician teachers and other exemplars recognized for 'demonstrated excellence in clinical care, leadership, compassion and dedication to service'" [3].

1. Schweitzer A. From a letter to his students,1953. Quoted in: Ann Intern Med. 1997;127:853.
2. Alpha Omega Alpha. Available at: http://www.alphaomegaalpha.org.
3. The Arnold P. Gold Foundation. Available at: http://humanism-in-medicine.org/about-us/

The Compleat Physician

The Compleat Physician is one who is capable in all three dimensions: he is a competent practitioner; he is compassionate; and he is an educated man. To use the classical terminology, he combines techné with philanthropia and paideia. Few men can perform with perfection, even adequately, at all levels. We must repress the tendency to apotheosize our profession by expecting all physicians to excel in all three.

American educator and author Edmund D. Pellegrino (1920–2013) [1]

Pellegrino was one of modern America's leading medical philosophers, ethicists, and humanists. He served as president of The Catholic University of America and as Chairman of the President's Council on Bioethics. He was a prolific author of scholarly articles and books, and founder of the Edmund D. Pellegrino Center for Clinical Bioethics at Georgetown University.

Fig. 2.12 (a) Kos, the island of Hippocrates. (b) Stamp with the likeness of Hippocrates, issued on the occasion of the unification of the island with the Greek motherland (1948). http://www.springerimages.com/Images/MedicineAndPublicHealth/1-10.1007_s11789-010-0014-y-2

Here I choose to focus on the third of the dimensions described by Pellegrino: *paideia*, being an educated human being. Fortunately, American physicians enter medical school only after some degree of liberal arts education, perhaps introducing them to literature, music, and visual arts, as they hurry toward acceptance in medical school. I point this out because, in many countries, medical school begins immediately following secondary education, and thus the graduating physician is quite focused on doctoring, without the leavening of a broad-based education.

Osler considered it important that young physicians continue to read widely outside medicine. His book, *Aequanimitas*, ends with a curiously unnumbered page with the heading: "Bed-side Library for Medical Students." Among Osler's recommendations are the Old and New Testament, Shakespeare, Marcus Aurelius, Cervantes' *Don Quixote*, and Emerson [2].

But what about after medical school and residency? Medical practice is demanding and time-consuming. And, of course, so is family life, especially if there are young children. And television and the computer are highly seductive. The early practice years are the time to establish what—I hope—will become lifelong habits of reading for pleasure, of attending cultural events, and of travel to the sites, such as Greece and Rome, where history occurred. Read some of Osler's recommendations. Go visit the Greek island of Kos and walk the ground where Hippocrates walked (see Fig. 2.12).

After a few years in small town practice, a member of the town board came to my office. He strongly encouraged me to become a member of a committee reporting to the town board. When I hesitated—after all, I was still building a busy practice—he pointed out, "Dr. Taylor, don't you realize? You are perhaps the most educated person in town. We need you to serve." Of course, with that comment, I felt I must join the committee.

That conversation also highlighted my—the physician's—duty to give back to the community, discussed next.

1. Pellegrino ED. Humanism and the physician. Knoxville TN: U Tenn Press;1979, p. 157.
2. Osler W. Aequanimitas with other addresses. Ed. 3. Philadelphia: Blakiston; 1932.

Giving Back to the World

There can be fewer luckier people in the world than the surgeons of India who have the opportunity to treat not only the rich but also the poor and destitute.

Roman Catholic nun Mother Teresa (1910–1997) [1]

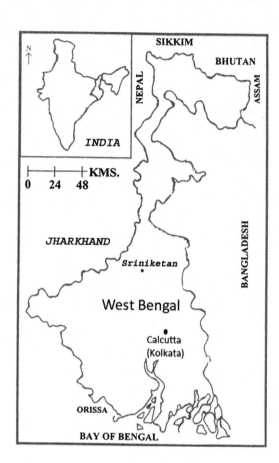

Fig. 2.13 The location of Calcutta in West Bengal, India. http://www.springerimages.com/Images/Environment/1-10.1007_s10661-012-2744-4-0

Sometimes called *Blessed Teresa of Calcutta* and best known for her work in the slums in the capital city of the Indian state of West Bengal, Mother Teresa was the founder of the Missionaries of Charity, which administers schools, orphanages, shelters, and hospices for the sick and needy. She received the 1979 Nobel Peace prize and was beatified in 2003. Her "mighty purpose" was service to the poor, sick, and disabled (see Fig. 2.13).

As we think about Mother Mary Teresa a few pages after a message from Dr. Albert Schweitzer, it is hard not to see the parallels in their lives. Although their lifetimes overlapped by 32 years, I find no evidence that they ever met. Yet both championed selfless care of the most needy; both achieved international recognition for their work; both received the Nobel Peace Prize (Schweitzer, 1952; Mother Teresa, 1979); both directly changed many lives for the better and inspired countless others to do the same.

Writing about "giving back to the world" in a recent issue of the *Journal of the American Medical Association*, Pescovitz invokes a metaphor that involves the Jordan River that feeds both the Sea of Galilee and the Dead Sea. The Sea of Galilee gives back the water to the Jordan River, which then flows south to the Dead Sea, where the current ceases. Fish and vegetation thrive in the Sea of Galilee while the waters of the Dead Sea are devoid of life—dead. The author summarizes: "In life, we all should be like the Sea of Galilee. We should give back to the world that gives us so much" [2].

What does this have to do with medicine today? Despite murmurings that today's medical students lack altruism, Streed describes a survey of 900 medical students at ten US medical schools, finding that 86 % believed in the concept of health care for all, and that two thirds would donate future time and income to support that care [3]. The surge of interest among medical students to help in disasters such as the Haitian earthquake or Hurricane Katrina may be an indication of a rising commitment to "give back."

1. Mother Teresa: Speech to the International College of Surgeons, Calcutta, 1998: Quoted in: McDonald P. The Oxford dictionary of medical quotations. New York: Oxford University Press; 2004, page 71.
2. Pescovitz OH. Swimming in the Sea of Galilee. JAMA. 2013;309:885.
3. Streed C. Where does our altruism go? Available at: http://boards.medscape.com/forums?128@921.yHFbagueebo@.2a382796!comment=1.

The Intimate Chambers of Our Patients' Lives

> As physicians, we are invited into the most intimate chambers of our patients' lives. We should acknowledge that unfettered trust with dignity, deference, and respect. For a physician, caring for patients is not only a duty; it is a privilege. Alleviating pain and restoring health for another human being induce an exhilaration few others experience in their careers.
>
> American heart surgeon Michael E. DeBakey (1908–2008) [1]

Michael DeBakey was a remarkable man, and not only because he lived just a few months short of 100 years. During a 75-year career in medicine, he operated on more than 50,000 persons; he invented the roller pump that was a key feature of the early heart-lung machines; and he was the first to use a successful external heart pump in a patient. In 2005 he suffered an aortic dissection, and his life was saved as

Fig. 2.14 Michael E. DeBakey. http://www.springerimages.com/search.aspx?caption=MICHAEL DEBAKEY

his team performed a *DeBakey Procedure*, a technique that DeBakey had developed [2] (see Fig. 2.14).

Despite DeBakey's larger-than-life accomplishments and his innovative thinking, he was—after all—a high-volume, operating-room-dwelling, cardiac surgeon. Thus I find it refreshingly curious that the phrase "we are invited into the most intimate chambers of our patients' lives" comes from a surgeon. I do not refer to the use of the word "chambers," as in heart chambers; this is a word heart surgeons use often. No, I am intrigued that this internationally renowned surgeon was sensitive to the personal vulnerability of his patients. This is an attribute stereotypically associated with family physicians, pediatricians, nurses, maybe psychiatrists, not necessarily surgeons.

As a medical school professor privileged to teach medical students throughout all 4 years of the curriculum, I encourage incoming doctors-to-be, garbed self-consciously in their new white coats, as follows: Not long ago you were college students, or perhaps working to earn some tuition money. Today you are *student doctors*. You will encounter patients in the office and hospital, often alone without a mentor to guide you. These patients will tell you secrets they would not tell their mothers. They will show you their bodies. They will let you touch them and sometimes invade various orifices. They will tell you what really worries them, which will sometimes come as a surprise. They deserve your respect, and the best you have to offer.

My wish is that all young doctors will finish their careers with honor and dignity, and with respect for their patients, as DeBakey did. If this happens, then the next generation of new physicians will inherit the sense of patient trust that we veteran healers did, as their first patients invite them into the intimate chambers of their lives.

1. Personal essay by Michael E. DeBakey, MD. In: Manning PR, DeBakey L. Medicine: preserving the passion, ed. 2. New York: Springer; 2004, p. 39.
2. DeBakey ME et al. Surgical considerations in the treatment of aneurysms of the thoracoabdominal aorta. Ann Surg. 1965;162:650.

Chapter 3
The Art and Science of Doctoring

In seeking resources for my books, I have spent many hours on line, following promising leads and a few dead-end paths. My journeys through the medical literature brought me to a message to young graduates by Dr. C.H. Low, published in the *Singapore Medical Journal* in 1998, admittedly not one of the publications I read regularly [1]. Dr. Low, crediting an anonymous source, offers this poem about doctoring:

I sought for happiness
Happiness I cannot see
I seek for richness of spirit
Richness eluded me
I looked for fulfillment
Fulfillment escaped me
But when I gave Medicine to my brethren
I found all three.

© Springer Science+Business Media New York 2015
R.B. Taylor, *On the Shoulders of Medicine's Giants*,
DOI 10.1007/978-1-4939-1335-0_3

Doctoring, the gerund form of the verb "to doctor," connotes the combination of competency in basic clinical abilities with training in communication and reasoning skills and proficiency in various psychosocial arenas such as mental health, domestic violence, and substance abuse. The components of "doctoring" mentioned here are the basis of a 4-year course in *Doctoring* pioneered at the University of California Los Angeles School of Medicine 20 years ago and now taught at a number of other institutions, sometimes under different course titles. According to Wilkes et al., although the course "has successfully taught large numbers of students psychosocial content and communications skills that are often overlooked in traditional medical school curricula and has had an impact on the larger culture of medical education, the authors believe that its full promise remains unfulfilled." They cite the greatest barriers as "cultural." After all, the *Doctoring* course aims to change the direction of medical education and encourage students "to think critically and to question fundamental aspects of the way medicine is taught, learned, and practiced" [2].

In fact, our current use of the word doctoring is bit of etymologic inconsistency, since "doctor" actually comes from the Latin *docere*, meaning "to teach," and it shares a common root with "docent," the knowledgeable person who conducts the museum tour. Also, there are doctors of economics, philosophy, law, and divinity. "Doctor" does not denote healing at all [3].

Nevertheless, doctoring as healing is what we physicians do when we are performing at our best, and striving to teach this approach to medical students and to employ it in our practices is well worth the effort. The words of some of medicine's giants in this chapter will offer some guidance as to the art and science of doctoring.

1. Low CH. Reflection for young doctors and doctors of tomorrow. Singapore Med J. 1998;39:535.
2. Wilkes MS, et al. The next generation of doctoring. Acad Med. 2013;88;438.
3. Taylor, RB. Medical wisdom and doctoring: the art of 21st century practice. New York: Springer; 2010, page 9.

More Than Technical Skill

Medical practice is not knitting and weaving and the labor of the hands, but it must be inspired with soul and be filled with understanding and equipped with the gift of keen observation; these together with accurate scientific knowledge are the indispensable requisites for proficient medical practice.

Physician and theologian Moses ben Maimonides (1138–1204) [1]

Born to Sephardic Jewish parents in Cordova, Spain, at a time when that country was under Moorish control, Maimonides and his family fled persecution in 1160, and migrated across northern Africa to the Holy Land, and eventually to Cairo, Egypt. During his career as a theologian and philosopher, he authored the *Mishneh Torah*, a 14-volume systemization of Talmudic law. As a healer, he was appointed

Fig. 3.1 Moses ben Maimonides. http://www. springerimages.com/Images/ MedicineAndPublicHealth/ 1-10.1007_978-1-4419-1034-9_3-2

court physician to the Sultan Saladin and the royal family of Egypt. As a medical writer, he authored treatises on asthma, hemorrhoids, fits (seizures), poisons and their antidotes, as well as *Regimen of Health*, a guide to healthy living (see Fig. 3.1).

Just as medical practice—doctoring—is not "knitting and weaving and the labor of the hands," it is also much more than cutting and sewing, diagnosing and treating. Yes, good healers are keen observers, an approach espoused by Hippocrates. They also, of course, have current, data-based scientific knowledge. But the very best healers bring more to the patient-physician encounter. How shall we describe this special trait, which Maimonides holds is inspired with the soul?

Pellegrino, as described in the previous chapter, would use the classical term *philanthropia*, describing compassion for a fellow human being [2]. Caring about another human being, for a physician, might be simply sitting by the bedside when it becomes clear that medicine has nothing more to offer, making the home visit when the elderly patient has no access to transportation, seeing the sick person when there is no hope of a professional fee, or attending the funeral of your patient, as a "final visit."

And yet today, this special trait we seek in the best physicians is under siege, by the demands of seeing too many patients in too little time and the increasing complexity of visits as "cyber-chondriacal" patients come with printouts from the web and with requests for medicines they have seen advertised on television. Compassion fatigue is very real today. If this seems to be happening to you, Pfifferling et al. recommend learning how to care for yourself—physically and emotionally—and by self-reflection, "coming to terms with the anger, fear, and self-doubt that some physicians have suppressed since medical school" [3].

If you feel that you have lost some of your *philanthropia*, and that you have no heart left to give, it is time to recognize that you have lost some of "the indispensable requisites for proficient medical practice." And it might be time to become the patient yourself, to overcome compassion fatigue.

1. Maimonides. Quoted in: Bull Institute History Med. 1915;3.
2. Pellegrino ED. Humanism and the physician. Knoxville TN: Univ Tenn Press;1979, p. 157.
3. Pfifferling JH, et al. Overcoming compassion fatigue. Fam Pract Mgt. 2000;7(4):39.

About Offering Hope

Always hold out hope for the patient, even if the symptoms point to a fatal issue, for nature often brings things to pass which seem impossible to the surgeon.

French surgeon Ambroise Paré (1510–1590) [1]

Considered by many to be at least one of the fathers of surgery, Paré is best remembered today as the battlefield surgeon who—without meaning to do

Fig. 3.2 Ambroise Paré. http://www.springerimages. com/Images/MedicineAnd PublicHealth/1-10.1007_ s00381-008-0775-5-0

so—provided us with a reasonable model of the controlled clinical trial (see Fig. 3.2). Traditionally battlefield wounds were treated with cauterization and boiling oil—for which the soldier-patient paid. But in one historic instance, there was a group of war-wounded who could not pay. These were treated with an ointment containing oil of roses, egg yolk, and turpentine. To everyone's surprise those patients receiving the ointment fared much better than those receiving boiling oil and cautery [2].

Despite the serendipity of his unintended scientific experiment, Paré's most valuable contribution to modern surgery was the introduction of the surgical ligature—instead of cautery—in amputations of extremities during wartime.

But let us return to his quotation about instances of probable fatal outcome. Certainly an optimistic attitude on the part of the physician/surgeon can ease the anguish of the war-wounded. But hope is also a powerful panacea for those with serious, life-threatening illness. There is only one problem: The healer must believe the hopeful words. If not, the doctor is being untruthful, and the patient will know this.

Hope need not always connote *cure*. Oncology specialists know this very well. Sometimes hope is for relief of pain, for the simple ability to enjoy one more meal without nausea, or to avoid a urinary catheterization. Often, being a good healer means being willing to try one more thing, and yet one more if that fails. At times it is allaying the fear of abandonment, even the "premature declaration of death," as friends avoid contact with a dying patient. And sometimes offering hope at the end of life means giving emotional support, just as Ivan Ilyich's young butler, Gerasim, came to the bedside, touched him, and gave him comfort [3].

I conclude by returning to Ambroise Paré, who seems, in his way, to have been a humble man. His most enduring saying was "I dressed him and God healed him" [4]. ←

1. Bloch H. Ambroise Paré: father of surgery as art and science. So Med J; 1991;84:763.
2. Ross JB et al. Ambroise Paré: a surgeon in the field. New York: Viking Penguin; 1981, page 558.
3. Tolstoy L. The death of Ivan Ilyich. In: Guerney BG, Ed. A treasury of Russian literature. New York: Vanguard; 1943, page 530.
4. Delacomptée JM, *Ambroise Paré, La main savant*. Paris: Gallimard; 2007, page 166.

Practical Medicine in Medical Practice

Practical medicine is never the same as scientific medicine but rather, even in the hands of
the greatest master, an application of it.

German pathologist and anthropologist Rudolph Virchow (1821–1902) [1]

Cited previously in Chap. 2, Rudolph Virchow is remembered for many scientific
accomplishments, including his contributions to cell theory of his day. He was the
first to identify leukemia cells under the microscope. His recognition that an enlarged
left supraclavicular node—aka "the seat of the devil"—can be the first sign of gastric
malignancy earned him eponymous immortality: the Virchow node (see Fig. 3.3).

I confess that I puzzled over the Virchow quotation above, reminding myself that
he was, fundamentally, a pathologist, a physician on the scientific side of the bell

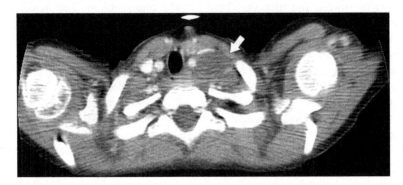

Fig. 3.3 Left supraclavicular (Virchow) node, see arrow. http://www.springerimages.com/Images/
MedicineAndPublicHealth/1-10.1007_978-1-4614-4872-3_10-16

curve. On the other hand, and while never claiming to be a "master," I am a family physician, with more than a decade of solo rural practice experience, plus another 40 years in teaching roles. As a primary care physician, I have reduced a dislocated shoulder during a house call, I have removed many fishhooks from a variety of sites on the skin, and I have treated anaphylaxis, hyperventilation, and fractured bones in my office. Yes, I practice "practical medicine" while doing my best to apply the latest in scientific medicine.

What Virchow was saying must be taken in the context of his time. Dutch physician Herman Boerhaave left his professional heirs a mysterious book—embellished in silver and locked—titled *Macrobiotic: The True and Complete Secret of Long Healthy Life* that was auctioned following his death in 1738 and sold for a high price. The excited buyer found every page blank but one, which states: "Keep the head cool, the feet warm, and the bowels open." Dechmann describes this as a "posthumous pleasantry" [2]. But Clark explains, "This legacy of Boerhaave to suffering humanity typified, not inaptly or unjustly, the acquirements, not of medical science, but of medical art at the close of the eighteenth century. Empiricism, authority, and theory ruled the medical practice of the world at this time" [3].

Certainly, Boerhaave's sardonic legacy was a call for a more logical approach to the care of patients, and Virchow's writing a century later—during the era of Semmelweis, Snow, and Pasteur—marked another milestone along the path to today's translational research and clinical practice built on a foundation of science.

1. Virchow R. Disease, life, and man: selected essays, Rather LJ, trans. Stanford CA: Stanford Univ Press; 1958.
2. Dechmann L. Dare to be healthy, ed 2. Seattle WA: Washington Printing Company; 1919. Available at: http://www.gutenberg.org/files/14985/14985-h/14985-h.htm.
3. Clarke EH. Practical medicine. Am J Med Sci. 1876;71:127.

Reconciling Uncertainty and Probability

Medicine is a science of uncertainty and an art of probability.

Canadian-American physician Sir William Osler (1849–1919) [1]

Osler was certainly medicine's all-time leading aphorist. In fact, he once described aphorisms as "burrs that stick in the memory," a saying that, itself, seems to qualify as an aphorism [1]. Of all human endeavors, probably none has produced more useful aphorisms than medicine. In fact, this book contains some of the best, and Osler contributed more than his share.

In the quote above, which links uncertainty and probability, Osler brings together the two faces of medicine—the science and the art—in ironic juxtaposition, something like describing Washington DC as a combination of Northern charm and

Fig. 3.4 Structure of double-helical DNA. http://www.springerimages.com/Images/LifeSciences/1-10.1007_978-1-4419-6324-6_18-4

Southern efficiency. Science, of course, is about the objective, the quantifiable, and the statistical probability that a hypothesis is supported or not. Art is about the craft of dealing with patients, about clinical experience, about creativity, and about how we as physicians deal with ambiguity. Science is about the determination of the effectiveness of a drug, such as the new diabetes medication canagliflozin (Invokana); art is about prescribing the drug for a patient, in the context of possible side effects, drug interactions, and all the nondiabetic entries on the patient's problem list.

In the 1970s there was enthusiasm for prospective medicine, a system of anticipatory health planning, in contrast to "reactive medicine." Prospective medicine attempted to help reconcile uncertainty and probability [2]. The cornerstone of prospective medicine was the Health Hazard Appraisal, an assessment of the patient's characteristics and behaviors in the context of what we know about various diseases; the outcome is a determination of 10-year risks as a tool to guide "health hazard" reduction [3].

Prospective medicine never really caught on, and we have not heard much about it for some four decades. But we may be seeing a resurgence of interest in anticipatory health care. Over the years we have developed more sophisticated methods of determining human biomarkers and disease risk. Currently, the major catalyst is the mapping of the human genome, completed in 2003, bringing the double helix to life [4] (see Fig. 3.4).

Genomic research is predicted to "drive the discovery of predictive factors and personalized medicine," according to Snyderman et al. Clinical, demographic, and family history data will now be supplemented by genotypic data, collectible at birth, that can have much better risk assessment for diseases such as breast cancer than was available in the last century [5].

Perhaps, with the use of genomic research, the science of uncertainty and the art of probability have moved a little closer together.

1. Osler W. Quoted in Bean WB. Sir William Osler: Aphorisms. New York: Henry Schuman, Inc.; 1951.
2. Robbins LC et al. How to practice prospective medicine. Indianapolis: Methodist Hospital of Indiana. 1970.
3. Robbins LC et al. Prospective medicine and the health hazard appraisal. In: Health Promotion: principles and clinical applications, Taylor RB, ed. New York: Appleton-Century-Crofts; 1982, Chap. 4.
4. Mistell T. The cell biology of genomes: bringing the double helix to life. Cell. 2013;152:1209.
5. Snyderman R et al. Prospective health care: the second transformation of medicine. Genome Biology. 2006;7:104

The Confidence of the Healer

What is the secret of success? To inspire confidence … The doctor who possesses this gift can almost raise the dead. The doctor who does not possess it will have to submit to the calling-in of the colleague for consultation in a case of measles.

Swedish psychiatrist Axel Munthe (1857–1949) [1]

Munthe, author, philanthropist, and animal rights activist, is not as renowned as Virchow or Osler, but he brings to our attention a vital, but seldom discussed, feature of good doctoring: inspiring confidence. "Raise the dead," even when qualified with "almost," is admittedly hyperbolic. Yet, healing is enhanced when the patient has confidence in the healer, and there can be no patient confidence if the doctor lacks clinical self-assurance.

Early in my academic career, I was astounded when, at a multidisciplinary curriculum committee meeting at my medical school, a senior and venerated professor of medicine announced to the group, "We all live with the secret fear that our colleagues will discover the depth of our incompetence." I don't recall the context of his revelation, but I found it stunning at the time. Since then I have learned that the phenomenon he was describing is termed the *imposter syndrome*.

First described in women, but also found in men (as described above), the imposter syndrome describes a fundamental lack of confidence in one's abilities and an inability to accept credit for accomplishments. Life's successes are dismissed as good luck, fortuitous timing, or the largess of others. In short, those with the imposter syndrome believe themselves to be "frauds," and they hope that those around them do not find out [2]. The imposter syndrome can occur in any line of work [3]. In physicians, it is probably more common than realized.

Binenbaum et al. surveyed 210 medical and surgical residents at the University of Pennsylvania in 18 specialties regarding their confidence levels, and their perception of various factors affecting their subjective confidence. The authors report, "Mean confidence increased during internship from 12 to 32 (on a 1–100 scale) but remained in the 50s during residency for most specialties." Factors fostering self-assurance were independent decision-making opportunities and good backup support, while high patient volumes, long work hours, and abusive interactions were described as "poorly valued items." The authors reported similar findings for both medical and surgical residents [4].

Of all the specialties, surgery seems to be the one in which self-assurance of the physician is related to inspiring patient confidence. It takes a certain kind of courage to cut into the abdomen or chest of another human being. For this reason, I read with interest the report of a study of "our trainees' confidence" involving 4,136 surgery residents, administered at the time of the American Board of Surgery In-Service Training Examination. The authors characterized self-confidence as "an attitude that allows individuals to have a positive and realistic perception of themselves and their abilities." They found some interesting characteristics of those who felt less confident in their operating skills: They were more likely to be single, childless, female, and, curiously, residents at larger, university-based programs in the northeastern USA. "The larger the residency program as measured by the number of chief residents, the less likely its residents were to feel adequately skilled for their level of training" [5].

In the routine of daily practice and in more anxious settings where vital therapeutic decisions are needed, the overly timid physician can be a poor healer, as can the overconfident one. If you find that you seek many more consultations than your colleagues and that you have serious doubts about your abilities, then you just might have the "medical imposter syndrome."

1. Munthe A. Quoted in: Johnson WM: the modern doctor of the old school. New York: Macmillan; 1936, p. 139.
2. Clance PR et al. The impostor phenomenon among high achieving women: dynamics and therapeutic intervention. Psychotherapy Theory, Research, and Practice. 1978;15:241.
3. Dowd SB et al. Do you feel like an impostor? Health Care Superv. 1997;15:51.
4. Binenbaum G et al. The development of physician confidence during surgical and medical internship. Am J Surg. 2007;193:79.
5. Bucholz EM, et al. Our trainees' confidence: results from a national survey of 4136 US general surgery residents. Arch Surg. 2011;146:907.

Disease, Man, and the World

> A physician is obligated to consider more than a diseased organ, more even than the whole man—he must view the man in his world.

American neurosurgeon Harvey Williams Cushing (1869–1939) [1]

Cushing was more than America's brain surgery pathfinder. He first described hypercortisolism, now called Cushing disease, he received the 1926 Pulitzer Prize for his biography of Sir William Osler and the Lister Medal for his contribution to surgery, and in 1988 he was pictured on a 45-cent US postage stamp. Also, as the quotation above suggests, he was an expansive clinical thinker when it came to the interrelationships of disease and the world (see Fig. 3.5).

When I was at an impressionable age—about 16—I read a Ray Bradbury science fiction story titled *A Sound of Thunder*, first published in *Collier's* magazine in 1952. The story, which I still recall today, tells of an adventurer who, using a

Fig. 3.5 Harvey Cushing. http://www.springerimages.com/Images/MedicineAndPublicHealth/1-10.1007_s11102-012-0402-z-4

futuristic but somewhat erratic time machine, travels to the past. He is warned, "Do not leave the elevated pathway." He does so, of course, and accidentally steps on a butterfly. When he returns home from his time travel, the world has changed in some very dark ways—with grim attitudes and evil people in power. The changes are the evolutionary consequences of the death of the butterfly he crushed, and today the phenomenon is called the *butterfly effect* [2].

I hold that there is a "butterfly effect" in the world's natural systems, including health. Think of a *hierarchy of natural systems* that looks like this:

Universe
World
Nations
Communities
Person
Body systems
Organs
Tissues
Cells
Molecules
Atoms

A change, a "perturbation," at any level in the hierarchy affects all levels above and below. Consider what happens when islet cells in the pancreas fail, causing diabetes, which affects the person and his or her interaction in the community, resources at various levels are utilized for care, and to some small degree, the world and the universe are affected, as are body systems, organs, tissues, and so forth. Or what happens when there is an outbreak of a serious illness in, for example, Asia. Resources are directed to this area of need, and thus are not available for care of more routine diseases people elsewhere may experience, which results in some change in availability, however small or large, of drugs or vaccines, with the resultant effects on health of organs, tissues, and cells [3].

Just as everything that happens today has implications for the future—the butterfly effect—every disease must be considered in the context of the world and all that implies.

1. Cushing H. Quoted in: Dubos RJ. Man adapting. New Haven, Connecticut: Yale Univ Press; 1965, Ch. 12.
2. Bradbury R. A sound of thunder. Collier's magazine. 1958; June 28.
3. Taylor RB. Family—a systems approach. Am Fam Phys 1979;20:101.

Caring for the Patient

The good physician knows his patients through and through, and his knowledge is bought dearly. Time, sympathy and understanding must be lavishly dispensed, but the reward is to be found in that personal bond which forms the greatest satisfaction of the practice of medicine. One of the essential qualities of the clinician is interest in humanity, for the secret of the care of the patient is in caring for the patient.

American physician Francis Weld Peabody (1881–1927) [1]

"… the secret of the care of the patient is in caring for the patient." These words closed a lecture given by Doctor Francis W. Peabody to Harvard medical students

Fig. 3.6 An early view of Harvard Medical School (right). http://www.springerimages.com/ Images/MedicineAndPublicHealth/1-10.1007_s12682-011-0091-9-3

on October 21, 1925. The lecture was one of a series of late-afternoon teaching sessions and subsequently published in the *Journal of the American Medical Association* (JAMA) in 1927. The title of the lecture and the paper was simply "The Care of the Patient."

A graduate of Harvard College and Harvard Medical School, Peabody pursued rigorous training at John Hopkins Hospital, in Germany and Denmark, and at the Rockefeller Institute in New York City. When the Peter Bent Brigham Hospital opened in 1913, Peabody was the first chief resident physician. He went on to an illustrious career, as a clinician, teacher, and humanist. Today at Harvard Medical School, medical students may join the Francis Weld Peabody Society (see Fig. 3.6).

The backstory is that at the time of his memorable lecture, Peabody was aware that he had an incurable cancer. He died of a sarcoma in 1927 at age 46 [2–4].

In 2002, Adler asked: "The emotional investment required to construct a caring doctor-patient relationship can be justified on humane grounds. Can it also be justified as a direct physiologic intervention?" He concludes that the answer is yes, citing "correlation of indicators of autonomic activity" in persons in an empathic relationship, and also the reduced secretion of stress hormones when one is in a caring relationship. Adler calls the process *sociophysiology* [5].

It would seem that Peabody was correct on two levels: Caring for the patient is not only the right thing to do; it can also have favorable—that is, healing—physiologic effects.

1. Peabody FW. The care of the patient. JAMA. 1927;88:877.
2. Paul O. The caring physician: the life of Dr. Francis W. Peabody. Cambridge, Mass.: Harvard Univ Press; 1991.
3. Harvey AM. Francis Weld Peabody: the blending of general internal medicine and clinical science. The Pharos. 1981:44(3):6.
4. Tishier PV. The care of the patient: a living testimony to Francis Weld Peabody. The Pharos. 1992;55(3):32.
5. Adler HM. The sociophysiology of caring in the doctor-patient relationship. J Gen Int Med. 2002;17:883.

Satisfaction and Choices in Medicine

It's the humdrum, day-in, day-out, everyday work that is the real satisfaction of the practice of medicine; the million and a half patients a man has seen on his daily visits over a 40-year period of weekdays and Sundays that make up his life.

American physician and poet William Carlos Williams (1883–1963) [1]

William Carlos Williams, the poet-doctor, was a generalist physician who wrote Pulitzer-Prize-winning poetry. He also wrote novels, plays, essays, and short stories. Colgan writes that Williams "worked harder at being a writer than he did at being a physician" [2]. Of course, I am sure that many of his patients would have disagreed with this statement. Williams devoted his medical life to serving the residents of his hometown, Rutherford, a New Jersey town not far from downtown Manhattan.

Certainly, Williams was a complex man, living almost schizophrenically in two worlds—the clinical and the literary. How many practicing general physicians, those who make home visits, are friends with great writers such as Ezra Pound, Wallace Stevens, and Allen Ginsberg? Williams once wrote: "I'm a doctor all day and into the night; I write on the run, or when it's plenty dark, before going to bed" [3]. Perhaps this presumably rushed pace and the habit of writing when fatigued from the day's work provides some insight into William's use of "strong, wry, bemused language" [3].

Was Williams a physician whose avocation was writing poetry? Or was he a poet who supported his literary efforts by practicing medicine, for which he had been trained? How many physicians today suffer similar conflicts? Are many of today's doctors really yearning to be writers?

There have been many successful physician writers: Richard Seltzer, W. Somerset Maugham, Michael Crichton, Arthur Conan Doyle, and Oliver Sacks. Russian physician-author Anton Chekhov described his double life—"medicine is my lawful, wedded wife, and literature is my mistress" [4]. Reisman et al. describe a physician-writer's workshop for resident physicians led by physician-author Abraham Verghese [5].

Not all physicians choose writing as their mistresses. Avitzur describes the avocations of neurologists, naming names and hobbies such as music, photography, and painting [6]. Others are hobby-farmers, boaters, or aviators.

Some writers, like Seltzer, continued to practice medicine as well as write; others, like Crichton, abandoned medicine to focus on their literary craft. Today, there seems to me to be an increasing tendency of doctors to leave medicine to follow other paths; a physician friend in Oregon retired early to become a vintner. Another is working to make a profit at what was his hobby farm. Perhaps this trend reflects frustration with the current practice environment, or maybe society has finally recognized that physicians are highly educated, very bright people who might just choose a career path that does not include patient care.

My point here is this: Williams continued to serve both his patients and his readers, exhibiting extraordinary energy. We know his heart was in medicine, because only a physician could have penned the words describing the real satisfaction of medicine, not to mention a million and a half patient encounters in a lifetime. And only a writer could have expressed the thought with such a precise choice of imagery and with such a cadence to the prose.

1. Williams WC. The autobiography of William Carlos Williams. New York: Norton; 1967.
2. Martin LW. The prose of William Carlos Williams. Middletown, Connecticut: Wesleyan University Press; 1970.
3. Coles R. William Carlos Williams: a writing physician. JAMA. 1981;245:41.
4. McLellan MF. Literature and medicine: physician-writers. The Lancet. 1997;349(9051):564.
5. Reisman AB et al. The craft of writing: a physician-writer's workshop for resident physicians. J Gen Int Med. 206;21:1109.
6. Avitzur O. Neurologists and their avocations: integrating work, home and play. Neurology Today. 2005;5(8):42,46.

Touching, Listening, and Technology

> It is as though when he talks or listens to a patient, he is also touching them with his hands so as to be less likely to misunderstand and it is as though, when he is physically examining a patient, they were also conversing.
>
> British author John Berger (1926–) [1]

In the sentence quoted above, Berger is describing country doctor John Sassel, practicing in the 1960s, as the protagonist evolves from being young, impatient, and disease focused to eventually become a seasoned, empathic healer. Described by one reviewer as "the most important book about general practice ever written," the author reveals elements of good doctoring—being available, applying scientific evidence, and, most important, listening to the patient [2].

Fig. 3.7 Phalen maneuver (photo credit: Dr. Harry Gouvas). http://commons. wikimedia.org/wiki/ File:Phalen_maneuver.jpg

Today there is disagreement about the best way to assess the patient. Traditionalists hold that a well-elicited medical history is the best route to diagnosis and a properly performed physical examination, including sometimes testing using classical manifestations such as Courvoisier sign or the Phalen maneuver, is fundamental to good doctoring (see Fig. 3.7).

Others disagree, citing studies questioning the worth of time-honored clinical signs, describing how much time can be saved by abbreviating direct patient contact, and extolling the merits of new technology. Why waste time asking the patient about symptoms when the information can be elicited as digital responses? In the doctor's lunchroom, we hear, "One look is worth a dozen listens." And we read articles wondering in print if the stethoscope is obsolete today [3].

All of this, of course, distances the physician from the patient. It decreases conversation and the opportunities for *touch*, in all its connotations. Here is what Canadian physician Ian McWhinney (1926–2012) would say to those who decry traditional patient contact: "What price will we pay for practicing medicine without touching our patients? The physical examination is more than a search for clinical data; it is one way we express our care and respect for the patient's body. Touch reaches out to the loneliness and isolation of the sick. Gentle touch—stroking or handholding—calms and soothes the anxious and troubled mind. Touch can reach us emotionally in a way our other senses cannot" [4].

1. Berger J. A fortunate man: the story of a country doctor. New York: Random House; 1967.
2. Feder G. A fortunate man: still the most important book about general practice ever written. Br J Gen Pract. 2005;55:246.
3. Wilkins RL. Is the stethoscope on the verge of becoming obsolete? Resp Care. 2004;48:1488.
4. McWhinney IR. The doctor, the patient and the home: returning to our roots. J Am Board Fam Pract. 1997;10:430.

Revealing the Physician's Human Side

Our noble profession would be well served if the public could be made aware of its more human side. A home visit or even a telephone call after a patient has been informed he has cancer, being present during the time of dying or in the operating room when surgery is performed or attending a former patient's funeral service serve those purposes well. Acceptance of these responsibilities and the discharging of them are the characteristics of a complete physician.

American physician Theodore E. Woodward, MD (1914–2005) [1]

The words above are taken from Woodward's address to graduating students at the University of Maryland School of Medicine in 1983. The well-informed reader

Fig. 3.8 Theodore
E. Woodward. http://www.
springerimages.com/Images/
LifeSciences/1-10.1007_
978-1-4419-1108-7_14-1

will recall Dr. Woodward, once chairman of the Department of Internal Medicine at the University of Maryland School of Medicine, as the putative author of the what may be medicine's most famous aphorism: "If you hear hoof beats, think of horses and not zebras" [1]. I'll tell more about Woodward and this saying in Chap. 5 (see Fig. 3.8).

Young physicians should not underestimate the value of Woodward's advice to show a human side. As I began my first practice experience after clinical training and obligatory uniformed service time, a veteran local physician advised two things: "If going on a house call at night, get very good directions; things look different in the dark." And, "When one of your patients dies—and some will—be sure to attend the viewing, to pay your respects to the patient and family. You and they are all part of this community."

There is, of course, a practical value to being a humanistic physician: patient satisfaction. A study by Hauck et al. of 185 randomly selected patients revealed the following: "A positive association was found between perceived physician humanism and patient satisfaction" [2]. To me this connotes allowing time for some non-medical conversation: "What do you think of this weather?" "How are your kids?" "Are you taking a vacation this year?"

There is one time-honored, logistical way to show the physician's human side. When I think of Woodward's remarks above, I think of one phrase: *being there for the patient*. On the telephone, at the bedside, in the home, in the operating room, and, in the end, among the mourners. A friend and fellow family physician, William (Bill) Phillips has given us a couplet that sums up what is important [3]:

You can pretend to know, you can pretend to care,

But you can't pretend to *be there*.

1. Woodward TE. Quoted in Colgan R. Advice to the young physician. New York: Springer; 2009.
2. Hauck FR et al. Patient perceptions of humanism in physicians: effects on positive health behaviors. Fam Med. 1990;22:447.
3. Phillips WR et al. The domain of family practice: role, scope, and function. Fam Med. 2001;33:273.

Chapter 4
Health, Disease, Illness, and Death

Health, and specifically connoting *good health*, may be, with tongue in cheek, described as the failure to make a careful diagnosis. After all, a good diagnostician can find at least three abnormalities—physical or behavioral—in virtually any person. Think of overeating, male pattern baldness, occasional dyspepsia, performance anxiety, flat feet, and more.

A disease is a perturbation in a body organ or function that results, sooner or later, in specific symptoms and signs. A cough, for example, is a symptom; hepatomegaly is a physical sign. Lung cancer is a disease, and an especially nasty one, at that (see Fig. 4.1).

© Springer Science+Business Media New York 2015
R.B. Taylor, *On the Shoulders of Medicine's Giants*,
DOI 10.1007/978-1-4939-1335-0_4

Fig. 4.1 Right lower lobe
lung cancer specimen after
lobectomy. http://www.
springerimages.com/Images/
MedicineAndPublicHealth/2-
ACNCR01-03-26-013

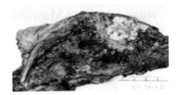

Illness describes both the disease and the individual's emotional and physical response—how the specific disorder affects the life of the ill person. Bronchogenic carcinoma typically results in progressive physical deterioration and eventual wasting leading to death. Some persons fight the inevitable end valiantly; others become despondent and withdrawn. And a few are saved by early detection and surgery, and may feel both elated and perhaps a little uncomfortable about being the lucky ones—so-called *survivor guilt*. Although the disease of lung cancer has some commonality at the tissue/organ level, the illness experience can vary from person to person.

Death, of course, is the end result of some disease, such as lung cancer, or injury; no one truly dies of old age. As one of my medical school attendings once observed, even the most needlessly worried hypochondriac will eventually die of some physical cause.

Disease and Its Manifestations

A disease also is further on the road to being cured when it breaks forth from concealment and manifests its power.

Roman philosopher and statesman Lucius Annaeus Seneca (4 BCE–AD 65) [1]

Sometimes called Seneca the Younger, the author of the words above was not only a philosopher but also a dramatist, statesman, and tutor to the Emperor Nero. The latter role was ill fated, however, when he was forced to commit suicide in the aftermath of a conspiracy to kill Nero. Despite probably being innocent of involvement, Seneca was obliged to open his veins and subsequently to take poison (see Fig. 4.2).

During his life, Seneca suffered serious health problems, which are reflected in his writings [2]. Perhaps his illnesses helped him to attain an insight into the nature of disease, and his words above remain relevant today. Simply stated, rational therapy, perhaps even curative therapy, cannot begin until there is a diagnosis.

Think about a young person with fever, fatigue, myalgia, joint pains, and headache. What disease does the patient have? The most common cause of this combination of symptoms is viral influenza. Infectious mononucleosis would also be on

Fig. 4.2 Lucius Annaeus Seneca. http://www.springerimages.com/Images/Geosciences/1-10.1007_1-4020-4497-6_120-32

the differential diagnosis list, especially if a sore throat were part of the picture. But consider that at least half of all patients with human immunodeficiency virus (HIV) infection initially present with just such symptoms, sometimes described as "the worst flu ever," occurring some 2–4 weeks following exposure to the virus [3]. With this acute constellation of findings, the HIV has begun to "manifest its power." At this time, the patient (and all the patient's sexual partners) will be fortunate if a prompt and accurate diagnosis is made, followed by timely therapeutic and public health decisions.

It is noteworthy that in Seneca's time, in the same century as Christ lived, there were actually few, if any, truly effective remedies. There were many recoveries, but few cures, at least in the true sense of the word. Today, we have many powerful drugs and almost-magical surgical procedures, and thus early recognition of disease manifestations is increasingly important. Hence, diagnostic delay becomes an issue.

In a report of 344 women with ovarian cancer, the median time between the onset of the most common symptom (abdominal pain) and diagnosis was 13 weeks [4]. A study of 180 men with testicular germ tumors revealed an average delay of 170 days from the onset of symptoms to diagnosis [5]. In many diseases, such as colorectal cancer, diagnostic delay can make a difference in outcomes [6].

Today, even more than in the time of Seneca, it is important to recognize when disease has broken forth from concealment and manifested its power in ways that can allow early diagnosis—and perhaps curative therapy.

1. Seneca LA. Moral epistles to Lucilius Junior, LVI. Paris: *Librairie Poussielgue Freres*; 1887.
2. Seneca LA. Quoted in: Edwards C. Death in ancient Rome. New Haven, Connecticut: Yale University Press; 2007, page 106.
3. Chu C et al. Diagnosis and initial management of acute HIV infection. Am Fam Phys. 2010;81:1239.
4. Tate AR. Determining the date of diagnosis—is it a simple matter? The impact of different approaches to dating diagnosis on estimates of delayed care for ovarian cancer in UK primary care. BMC Med Res Methodology. 2005;9:42.
5. Dieckman DF. Testicular tumors: presentation and role of diagnostic delay. Urol Int. 1987;42:241.
6. Tørring ML et al. Time to diagnosis and mortality in colorectal cancer: a cohort study in primary care. Br J Ca. 2011;104:934.

The Movement of Blood as a Disruptive Discovery

For the concept of the circuit of the blood does not destroy, but rather advances traditional medicine.

English physician William Harvey (1578–1657) [1]

Today high school students understand the work of the heart and the circulation of the blood. It was not always so. In William Harvey's time, medicine still clung to some beliefs that had evolved only modestly since the time of Aristotle (384 BCE– 322 BCE), who considered the heart a dry, hot, three-chambered organ that was the origin of the nerves and the seat of intelligence. Later, Galen (130–200) conceptualized the heart as a heat-producing furnace [2]. He taught that "nutritive blood" was

Fig. 4.3 William Harvey.
http://www.springerimages.
com/search.aspx?caption=
william harvey

produced in the liver and carried through veins to the organs, where it was consumed. "Vital blood," on the other hand, was made by the heart and pumped through arteries to carry the "vital spirits." In the sixteenth century the anatomist Vesalius (1541–1564) advanced our thinking regarding the heart's structure, but its function was still unclear.

Then came Harvey in the seventeenth century, writing: "It may very well happen thus in the body with the movement of the blood. All parts may be nourished, warmed and activated by the hotter, perfect vaporous, spirituous and, so to speak, nutritious blood. On the other hand, in parts the blood may be cooled, thickened, and be figuratively worn out. From such parts it returns to its starting point, namely the heart, as if to its source or to the center of the body's economy, to be restored to its erstwhile state of perfection. Therein, by the natural, powerful, fiery heat, a sort of store of life, it is reliquified and becomes impregnated with spirits and (if I may so style it) sweetness. From the heart it is redistributed. And all these happenings are dependent on the pulsatile movement of the heart" [1] (see Fig. 4.3).

The circulatory system described in Harvey's 72-page book was a disruptive discovery in its day. In the world of medical science, it may be compared to the discoveries that the world is round and not flat, and that earth (and other planets) orbit around the sun, and not the other way around.

In patient care, we now know that you can give too much oxygen to premature infants, that immunizations do not cause developmental disorders in children, and that high blood pressure is not a normal phenomenon with aging. How many more disruptive discoveries will occur during your lifetime and mine? And who will make them?

1. Harvey W. An anatomical exercise on the motion of the heart and blood in living beings. Frankfurt, Germany: *Sumptibus Guilielmi Fitzeri*; 1628.
2. Mowry B. From Galen's theory to William Harvey's theory: a case study in the rationality of scientific theory change. Stud Hist Philosoph Sci. 1985:16:49.
3. Aird WC. Discovery of the cardiovascular system: From Galen to William Harvey. J Thromb Haemost. 2011;9 Suppl 1:118.

Describing the Pain of Gout

The patient goes to bed and sleeps quietly till about two in the morning, when he is awakened by a pain which usually seizes the great toe, but sometimes the heel, the calf of the leg or the ankle. The pain resembles that of a dislocated bone … and this is immediately succeeded by a chillness, shivering, and a slight fever. [The pain] grows gradually more violent every hour, and comes to its height towards evening, adapting itself to the numerous bones of the tarsus and metatarsus, the ligaments whereof it affects; sometimes the gnawing of a dog, and sometimes a weight and constriction of the parts affected, which become so exquisitely painful as not to endure the weight of the clothes nor the shaking of the room from a person's walking briskly therein. The night is spent in torture.

English physician Thomas Sydenham (1624–1689) [1]

I urge you to admire the elegance with which Sydenham, sometimes called the "English Hippocrates," describes the symptoms of gout. This poetic description betrays that the author was a victim of the disease (see Fig. 4.4). (In fact, in the history of the Western world, gout seems to have sought out some of our most illustrious ancestors: Alexander the Great, John Calvin, Charlemagne, Charles Dickens,

Fig. 4.4 Thomas Sydenham. http://www.springerimages. com/Images/MedicineAnd PublicHealth/1-10.1007_978- 1-4614-0170-4_1-10

Fig. 4.5 Acute gout of the great toe. http://www.springerimages.com/Images/MedicineAndPublicHealth/1-10.1007_s00108-006-1578-y-1

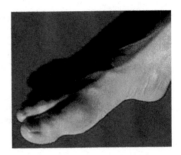

Benjamin Franklin, Isaac Newton, Peter Paul Rubens, Alfred Lloyd Tennyson, and Queen Victoria.) But consider the prose. Today's medical reference sources simply do not contain the vivid imagery found in many of the classical descriptions of disease.

Consider, for example, the following factually correct but rather colorless description of gout symptoms: "Gout attacks cause sudden severe joint pain, often with redness, swelling, and tenderness of the joint. Although an attack typically affects a single joint, some people develop a few inflamed joints at the same time. The pain and inflammation are worst within several hours and generally improve completely over a few days to several weeks, even if untreated" [2] (see Fig. 4.5).

There seems to be little place in current medical journals for literary flair [3]. But why not, sometimes, include lively disease descriptions in reference books, and delight the reader while also educating?

1. Sydenham T. A treatise on gout and dropsy, 1683. In: Works: A treatise on gout and dropsy. London: Robinson, Otridge, Hayes, and Newbery. 1788.
2. Gout symptoms: In: UpToDate. Available at: http://www.uptodate.com/contents/gout-beyond-the-basics.
3. Taylor RB. Medical writing: a guide for clinicians, educators and researchers, 2nd edition. New York: Springer: 2011, pages 265–287.

Scurvy, Citrus, and History

The first indication of the approach of the disease is generally a change of color in the
face, from the natural and usual look, to a pale and bloated complexion; with a listless-
ness to action, or an aversion to any sort of exercise. When we examine narrowly the
lips, or the caruncles of the eye, where the blood vessels lie most exposed, they appear
of a greenish cast. Meanwhile the person eats and drinks heartily, and seems in perfect
health; except that his countenance and lazy inactive disposition portend a future
scurvy.

Naval physician James Lind (1716–1794) [1]

A few pages ago, I wrote of William Harvey's description of the circulation of
the blood as a disruptive discovery. James Lind's recognition of the favorable influence of fruits on the prevention and treatment of scurvy was both disruptive in its
concept and epic in its influence on the world of seafaring. And, in its day, on naval
warfare.

We rarely see scurvy today, and as a US family physician for more than 50 years,
I must say that if I ever encountered a case in this country, I missed it. It was not
always so, especially when it came to seamen on long voyages.

Before the time of Lind, naval ships were obliged to return to port about every
10 weeks. According to Bollett, naval historians usually describe this as needing

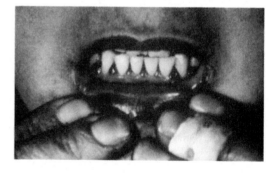

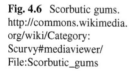

Fig. 4.6 Scorbutic gums.
http://commons.wikimedia.
org/wiki/Category:
Scurvy#mediaviewer/
File:Scorbutic_gums

to "refit" the ships, while the truth is that by 2–3 months at sea, the crew had become weakened by scurvy and needed to come home to recover [2].

As is common with a disruptive discovery, it took almost four decades before the Royal Navy fully implemented Lind's recommendation that seamen receive regular rations of oranges and lemons. According to Admiralty records, 1.6 million ounces of lemon juice was provided to naval seamen between 1795 and 1814. The practice earned them the sobriquet "Limeys."

Not needing to return to port for "refitting" effectively doubled the size of the British fleet, giving England a tactical advantage in naval conflicts. The health of the English sailors after weeks at sea compared with their French adversaries in the dawn of the nineteenth century was important in Nelson's victories over Napoleon's fleet that culminated in the battle of Trafalgar in 1805 [2]. Later, scurvy was endemic among soldiers in the forts in the American West; it was the most common disease reported from the frontier posts. In the US Civil War, scurvy occurred in many soldiers on both sides of the conflict [2] (see Fig. 4.6).

Near the end of the nineteenth century, infantile scurvy was noted. It happened as women ceased breast-feeding, and nourished their infants with vitamin C-deficient boiled cow's or "condensed" milk. Why did they do this? One reason was to avoid the possibility of transmitting tuberculosis—a fearful plague of the time—to offspring. Another reason was that, for the affluent, breast-feeding was somehow unseemly, resulting in the paradoxical finding of infant scurvy among children of the well-to-do [2].

What about today? In 2013 Congresswoman Suzan DelBene reported living, for a short time, on a food-stamp budget, estimated by the US government to be about $30 per week or about $4.50 per day. Her dietary staples during this time: oatmeal, macaroni and cheese, and peanut butter and jelly sandwiches. I don't see much in the way of fresh fruits in this diet: no oranges, lemons, or even limes. With 47 million Americans receiving food stamps, might we just have some cases of scurvy right here in modern America? [3].

1. Lind J. A treatise on the scurvy. Edinburgh; 1753.
2. Bollett AJ. Plagues and poxes. New York: Demos; 2004, pages 173–189.
3. Rep. DelBene eating on food stamp budget for week. HeraldNet. Available at: http://www.heraldnet.com/article/20130619/NEWS01/706199925.

On Precise Names of Diseases

Clear and precise definitions of diseases, and the application of such names to them are as expressive of their true nature, are of more consequence than they are generally imagined to be. Untrue or imperfect ones occasion false ideas; and false ideas are generally followed by erroneous practice.

English surgeon Percival Pott (1714–1788) [1]

Here I choose to reflect on one of the ironies in medical history. Pott, a pioneer in orthopedic surgery, was a strong advocate of precision in medical discourse and writing, calling for meticulous descriptions of disease and names that reflected the underlying pathology. This is slightly amusing, considering that "Pott's" has been as an eponym in the names of a number of diseases.

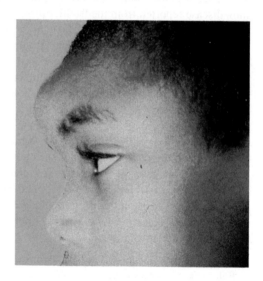

Fig. 4.7 Pott puffy tumor.
http://www.springerimages.
com/Images/
MedicineAndPublicHealth/2-
AID04E3-04-026B

Pott's disease is extrapulmonary tuberculosis of the spine, causing spondylitis and a deformity that has been found in the remains of mummies from ancient Egypt. Another disease named for Pott, who described it in 1760, is Pott puffy tumor (PPT), not exactly a paragon of "correct" nomenclatural. PPT is a rarely encountered subperiosteal abscess associated with osteomyelitis, sometimes seen in patients with frontal sinusitis (see Fig. 4.7).

Then there is the Pott fracture, a combined fracture and dislocation of the ankle. Medical lore holds that Pott himself suffered this fracture, but Gordon tells another story: Pott did indeed fall from his horse, but the injury suffered was actually an open fracture of the femur. Refusing to be taken to the hospital, he dispatched a servant to purchase a door. He was then placed on the door and transported home, where his leg was splinted, and eventually healed completely [2]. Legend holds that, in refusing to go to Saint Bartholomew Hospital, where he was an attending physician, Pott asserted that the only capable surgeon in that hospital (himself) was injured [3].

A few years ago, eponymic possessives began to come under attack. Pott's puffy tumor was changed to Pott puffy tumor. Down's syndrome became Down syndrome. Parkinson's disease is now Parkinson disease.

Today we have the campaign to eliminate eponyms altogether. Lou Gehrig disease must be called amyotrophic lateral sclerosis. No more Caffey disease; it must be infantile cortical hyperostosis. Crohn disease is now regional enteritis. But some diseases are holdouts: It takes just too many words to apply descriptive names to Hodgkin disease, Lyme disease, Raynaud disease, and, yes, Pott puffy tumor.

1. Pott P. A treatise on fistula in ano, 1767. In: Chirurgical works, vol III. London: Wood and Innes; 1808.
2. Gordon R. The alarming history of medicine. New York: St. Martin's Press; 1994.
3. Taylor R. White coat tales: medicine's heroes, heritage, and misadventures. New York: Springer; 2008, page 189.

The Beginning of the End of Smallpox

I hope that some day the practice of producing cowpox in human beings will spread over the world—when that day comes, there will be no more smallpox.

English general practitioner Edward Jenner (1749–1823) [1]

All physicians are—or should be—familiar with the story of how country doctor Edward Jenner knew of milkmaids who had contracted cowpox and who seemed somehow to be resistant to smallpox (see Fig. 4.8). In 1796 Jenner theorized that there was some sort of protection in the cowpox pus, and tested his hypothesis by injecting cowpox material into the arm of a local boy. Later he challenged the boy with smallpox, but the boy remained healthy. In 1798, Jenner published his monograph, *Inquiry into the Causes and Effects of the Variolae Vaccine*. Jenner also coined the word "vaccination," from the Latin word *vacca*, meaning cow. And for etymologic purists, "vaccine" and "vaccinate" should really only be used in connection with materials that come from cows, of which the cowpox virus is the only one I know.

Despite initial criticism and resistance—think of injecting healthy people with material from a diseased animal, horrors!—common sense won out. Jenner, like

Fig. 4.8 Edward Jenner (Public domain). http://commons.wikimedia.org/wiki/Edward_Jenner#mediaviewer/File:Edward_Jenner.jpg

Salk and Sabin in later years, lived to see the widespread benefits of his discovery. He is quoted as reflecting: "The joy I felt as the prospect before me of being the instrument destined to take away from the world one of its greatest calamities (smallpox) was so excessive that I found myself in a kind of reverie" [2]. Jenner's suggestion that someday there may be no smallpox must have seemed fanciful at the time. After all, in the eighteenth century, smallpox killed 400,000 annually in Europe, and left survivors with disfiguring scars or blindness [3]. Yet, this is just what happened.

Jenner was widely honored for his achievements, and yet he sought no personal enrichment. In fact, his crusade for widespread vaccination caused his country practice to suffer. His reward was to see vaccination replace variolation as the world's standard for smallpox prevention [3].

In 1967, the World Health Organization (WHO) set out to eradicate smallpox. In 1977, the last reported case of smallpox occurred in an unvaccinated hospital cook in Somalia. After a period of expectant observation, in 1980 the WHO declared smallpox an eradicated disease [4].

According to the latest information I can find, there are still existing samples of the smallpox virus: "451 at the Centers for Disease Control and Prevention in the United States and about 120 stored in a lab called 'Vector' in a remote Siberian town in Russia—continue to hibernate in liquid nitrogen." This is in spite of a 1996 decision to destroy these samples [4]. I don't know about you, but in a world where we long ago ceased to vaccinate our children, and where we look over our shoulders for possible extremist terrorists, I find the existence of this virus almost as frightening as the nuclear aspirations of some unfriendly nations.

1. A portrait of Dr. Edward Jenner. Barbenpd6. Available at: http://barbenpd6.wikispaces.com/Edward+Jenner.
2. Jenner E. Quoted in: Baron J. Life of Edward Jenner. London: Henry Colborn; 1838; Vol 1, Chap. 4.
3. Riedel S. Edward Jenner and the history of smallpox and vaccination. Proc Bayl Univ Med Cent. 2005;18:21.
4. Edward Jenner: the smallpox vaccine. Available at: http://perryjgreenbaum.blogspot.com/2011/03/edward-jenner-smallpox-vaccine.html.

About the Great Pox

I may be told by some that men may contract syphilis by sitting in a public privy; to this I can only answer that I have never witnessed a single instance; nor did the late Mr. Obre, who had been for many years most extensively engaged in treating the venereal disease; for on asking him if he believed the disease was propagated in this manner, he shrewdly answered, that it sometimes was the manner in which married men contracted it, but unmarried men never caught it in this manner.

Irish surgeon Abraham Colles (1773–1843) [1]

When we think of Colles, we think first of the fracture of the distal radius that bears his name: Colles fracture. He described this injury in his paper *On the Fracture of the Carpal Extremity of the Radius*, written some 80 years before X-rays were

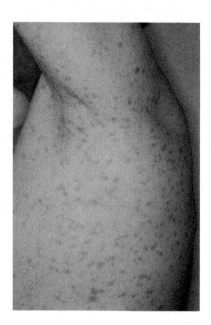

Fig. 4.9 Maculopapular rash of secondary syphilis. http://www.springerimages.com/Images/MedicineAnd PublicHealth/2-AID05E3-09-030

discovered. He also wrote a two-volume textbook titled *Lectures on the Theory and Practice of Surgery*, and the monograph on syphilis from which the quote above was borrowed [1].

I included Colles' observation about the ways in which syphilis might be contracted, not because these comments greatly advanced our knowledge of the disease and not because syphilis was his "signature topic." I wanted a little disease humor in the book, and also because the narrative struck a familiar chord: In 1959, when I was a sophomore medical student at Temple University School of Medicine in Philadelphia, we had our one and only lecture by Dr. John A. Kolmer (1886–1962), a renowned syphilologist and the "father" of the Kolmer test for the disease. Here is the story he told our class:

> A man came to me saying, "I caught syphilis in a public toilet."
> I replied, "That's an awful place to take a woman."

Syphilis can arguably be described as the most enigmatic and colorful of diseases. The disease takes its name from Sipylus, hero of a love poem titled *Metamorphoses*, written by the Roman poet Ovid (43 BCE–17 AD). The origins of the disease are obscure, but one theory is that it came to Europe as early explorers returned from the Americas. In the years that followed syphilis was often blamed on one or another ethnic group. At times it has been called the French disease, the Spanish disease, the Neapolitan disease, the Polish disease, and the Chinese pleasure disease. The Arabs have called it the Christian disease [2].

Syphilis, with skin lesions as one of its manifestations, was called the "great pox," to distinguish it from smallpox (see Fig. 4.9).

Unfortunately, the "great pox" is still with us. According to the US Centers for Disease Control and Prevention (CDC), there are 55,400 new cases of syphilis in the USA each year. "There were 46,042 reported new cases of syphilis in 2011, compared to 48,298 estimated new diagnoses of HIV infection and 321,849 cases of gonorrhea in 2011. Of new cases of syphilis, 13,970 cases were of primary and secondary (P&S) syphilis, the earliest and most infectious stages of syphilis" [3].

1. Colles A. Practical observation on the venereal disease, and on the use of mercury. Philadelphia: A. Walde; 1837, page 184.
2. Taylor RB. White coat tales: medicine's heroes, heritage, and misadventures. New York: Springer; 2010, page 39.
3. Syphilis CDC Fact Sheet. Available at: http://www.cdc.gov/std/syphilis/stdfact-syphilis.htm.

Great Outbreaks as Things of the Past

You and I may not live to see the day, and my name may be forgotten when it comes, but the time will arrive when great outbreaks of cholera will be things of the past; and it is the knowledge of the way in which the disease is propagated which will cause them to disappear.

British anesthesiologist and epidemiologist John Snow (1813–1858) [1]

Some of the medical giants in this book, such as Edward Jenner and James Lind, came from solid, but actually undistinguished, backgrounds, and became known by virtue of their discoveries. This was not the case with John Snow, considered by many the father of modern epidemiology. But even before his heroic actions in regard to the 1854 Broad Street cholera outbreak in London, he was a member of the Royal College of Physicians, a founding member of the Epidemiological Society of London, and the anesthesiologist who administered chloroform to Queen Victoria during the birth of her last two children (see Fig. 4.10).

The quotation above is a comment reportedly made in 1854 by Snow to the Reverend Henry Whitehead of St. Luke's Church in London's Soho district, a

Fig. 4.10 John Snow. http://
www.springerimages.com/
Images/Geosciences/1-10.
1007_978-94-007-1667-4_3-0

Fig. 4.11 John Snow's map
of cholera cases in 1854.
http://www.springerimages.
com/Images/MedicineAnd
PublicHealth/1-10.1007_978-
1-4419-6892-0_10-0

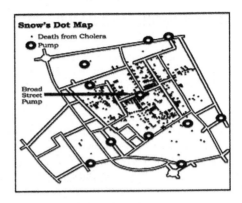

cleric who originally challenged the Broad Street pump theory as to the source of the cholera epidemic. He later came to agree with Snow. Snow carefully plotted the pattern of infection, shown in what Johnson calls the *ghost map*, eventually noting its epicenter at the now notorious Broad Street pump [1] (see Fig. 4.11).

Consider that the average life expectancy at birth in Snow's time was about 40 years, compared with a 2014 US estimate of 78.75 years [2]. But the almost-doubling of the life expectancy can be attributed, in great measure, to just what Snow observed: the knowledge of the way in which diseases are propagated and, subsequently, their prevention by hygienic methods that assure safe food and water, community immunization programs, and efforts to control epidemics promptly when they arise.

1. Snow J, 1854. Quoted in: Johnson S. The ghost map: the story of London's most terrifying epidemic—and how it changed science, cities, and the modern world. New York: Riverhead Books, 2006.
2. Life expectancy 2014: Available at: http://www.geoba.se/population.php?pc=world&type=015 &year=2014&st=rank&asde=&page=1.

Hand-Washing and Childbed Fever

I have assumed that the cadaveric material adhering to the examining hand of the accoucheur is the cause of the mortality in the First Obstetrical Clinic; I have eliminated this factor by the introduction of the chlorine-washings.

Hungarian physician Ignaz Semmelweis (1818–1865) [1]

Ignaz Semmelweis was a pioneer in many ways, not only in prevention of obstetrical infection, but also in the design of a clinical study and analysis of data. Briefly stated, he noted that among his hospital's two obstetrical divisions there was a big difference in mortality rates. Physicians and medical students, who routinely examined

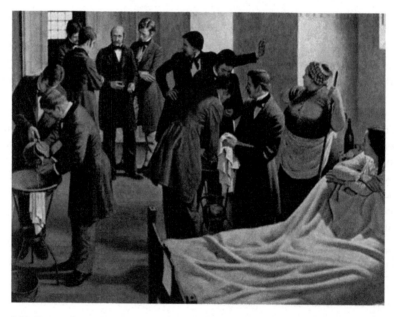

Fig. 4.12 Ignaz Semmelweis overseeing the washing of doctors' hands. http://www.springerimages.com/Images/MedicineAndPublicHealth/1-10.1007_s12262-012-0583-8-3

obstetrical patients shortly after having performed autopsies, and without interval hand-washing, staffed the First Obstetrical Division. The Second Division was staffed by midwives, who did not perform autopsies. The mortality rate in the First Division, three times that of the Second Division, dropped to half the rate of the Second Division after the initiation of hand-washing in 1847 [2] (see Fig. 4.12).

Encountering fierce resistance to his discovery—physicians couldn't really be the bearers of childbed fever—Semmelweis lobbied for his beliefs largely by writing letters to fellow obstetricians. It was not until 1861 that he published his now-classic treatise: *The Etiology, the Concept, and the Prophylaxis of Childbed Fever* [1].

Eventually his mind failed and he suffered early dementia. He was committed to an asylum. Shortly after his internment, he died at age 47 of an infection, presumably caused by the same sort of infection that his hand-washing technique prevented in countless young women.

In his monograph Semmelweis reflected: "When I look back upon the past, I can only dispel the sadness which falls upon me by gazing into that happy future when the infection (puerperal sepsis) will be banished…. The conviction that such a time must inevitably sooner or later arrive will cheer my dying hour" [1].

Today classic puerperal fever is uncommon, but this is not the case with surgical site and hospital acquired infections. Since Semmelweis' time, bacteria have proved to be wily adapters, and even with intensive routine hand washing and strict isolation protocols, between 5 % and 10 % of patients in US hospitals suffer hospital-acquired infections, resulting in numerous deaths and some $45 billion dollars in annual health care costs [3].

1. Semmelweis IP. The etiology, the concept, and the prophylaxis of childbed fever. Murphy FP, trans. London: Medical Classics; 1941, foreword. Original publication: Die Aetiologie, der Begriff und die Prophylaxis des Kindbettfiebvers. Pest, Vienna, and Liepzig: CA Hartleben Verlag; 1861.
2. Broemeling LD. Studies in the history of probability and statistics: Semmelweis and childbed fever. Available at: http://www.mdanderson.org/education-and-research/departments-programs-and-labs/departments-and-divisions/division-of-quantitative-sciences/research/biostats-utmdabtr 00504.pdf.
3. Krein SL. Preventing hospital-acquired infections: A national survey of practices reported by US hospitals in 2005 and 2009. J Gen Intern Med. 2012;27:773.

On Chorea, Dancing, and Genetic Screening

> Chorea is essentially a disease of the nervous system. The name "chorea" is given to the disease on account of the dancing propensities of those who are afflicted with it, and it is a very appropriate designation.

American physician George Huntington (1850–1916) [1]

The word chorea, also used in connection with Sydenham chorea, comes from an ancient Greek word describing a circle dance with singing that was mentioned in Homer's Iliad, written in approximately the eighth century BCE.

Huntington's chorea, now called Huntington disease (HD) because the disease has manifestations other than chorea, is an autosomal dominant disorder combining extrapyramidal choreiform movements with mood disorders and progressive

Fig. 4.13 George Huntington. http://commons.wikimedia.org/wiki/George_Huntington#mediaviewer/File:Georgehuntington.jpg

dementia. It chiefly affects persons between 20 and 50 years of age, destroying basal ganglia cells that are the seat of movement, emotion, and cognition (see Fig. 4.13).

Sometimes a disease becomes known by who has it. The famous person with Huntington disease was folk singer and composer Woody Guthrie (1912–1967), who wrote scores of popular songs including "This Land is Your Land." Consider this: We often call amyotrophic lateral sclerosis "Lou Gehrig disease." If Huntington disease had not already had an eponymic designation, might we remember it as "Woody Guthrie disease?"

Huntington, by following a family of persons with HD through several generations, was able to conclude, "When either or both the parents have shown manifestations of the disease…, one or more of the offspring almost invariably suffer from the disease…. But if by any chance these children go through life without it, the thread is broken and the grandchildren and great-grandchildren of the original shakers may rest assured that they are free from the disease." [1] This was long before the principles of Mendelian inheritance were elucidated in the early twentieth century.

A presymptomatic genetic screening test has been available for two decades. Because the first manifestations of disease typically occur after a person has had children, knowing if one carries the Huntington gene can be valuable in family planning as well as in anticipation of career and life events. Yet, curiously, Walker reports that only one in 20 persons at risk for HD has chosen to have the test. The chief reason cited is that no treatment is available [2].

Incidentally, for all physicians who wonder about the possible significance of their clinical observations, Huntington was a family physician, practicing with his father and grandfather in East Hampton on Long Island. The words I present above were the first sentences in the first paper Huntington ever published.

1. Huntington G. On chorea. Med Surg Reporter (Philadelphia). 1872;26:317.
2. Walker FO. Huntington's disease. The Lancet. 2007;369:218.

Disease and the Destiny of Humankind

Disease has played a dominant role in the destiny of the human race. It has destroyed old races and cleared the terrain for new ones which are hardier, more resourceful and more intelligent. It has defeated armies, paralyzed trade, and altered the economic life of nations. It has wiped out old castes and created new ones. It has destroyed explorers and colonizers, scattered their settlements and determined the ownership of continents…. Struck down by it with dramatic suddenness, great leaders, armies and nations, like ships without helmsmen, have crashed on the rocks.

American physician and author Ralph H. Major (1884–1970) [1]

Let's start with the fall of ancient Athens, as the city was weakened by a plague that swept the city from 430 to 426 BCE. The exact cause is debated, but one group of scientists holds that the pandemic was caused by typhoid [2]. Cartwright suggests that successive epidemics—malaria and bubonic plague—contributed to the fall of

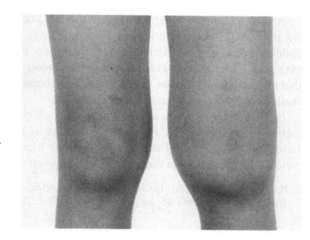

Fig. 4.14 Hemarthrosis after minor trauma in a boy with severe hemophilia A. http://www.springerimages.com/Images/MedicineAndPublicHealth/2-PEDIA01-08-020A

Rome in the fifth century [3]. Smallpox afflicting the indigenous population played a major role in the Spanish conquest of the Aztec nation in Mexico in 1519. In the 1770s King George III of England, who is thought to have had porphyria, treated the American colonists outrageously, helping to trigger the American Revolution. The outcome of Napoleon's adventure into Russia in 1812 was influenced more by "General Typhus" than by General Napoleon. Hemophilia so weakened the ruling Romanov dynasty that they were overwhelmed by the Bolsheviks in 1817 (see Fig. 4.14).

With the arrival of seafaring outsiders, measles, whooping cough, and influenza decimated the immunologically naïve Hawaiian people in 1848–1849. Scurvy was common among troops, especially confederate soldiers, during the American Civil War. Half of all deaths during WWI in Europe were caused by influenza.

Even diseases of plants can influence history. The potato blight of Ireland in the mid-nineteenth century prompted a massive migration to America, with subsequent influence on US demographics [4].

Malaria continues to cause three million deaths annually. More than 34 million people in the world have the disease, approximately two thirds of whom live in sub-Saharan Africa, where there are also large numbers of acquired immunodeficiency syndrome (AIDS) orphans. Sherman writes of malaria: "It's also a disease that is modern and yet has its parallels with the past in the kind of reactions that populations have when there's an unforeseen epidemic" [5].

When we add in the potential effects of *Filoviridae* viruses escaping from Africa and spreading by person-to-person transmission or the sporadic emergence of influenza virus variants related to farm animals and birds, we recognize that disease still shapes the destiny of humankind.

1. Major RH. Disease and destiny. New York: Appleton-Century;1936, foreword.
2. Typhoid may have caused fall of Athens, study finds. National Geographic. Available at: http://news.nationalgeographic.com/news/2006/02/0227_060227_athens_plague.html.
3. Cartwright FF. Disease and history. New York: Crowell; 1972, pages 29–53.
4. Billings M. The influenza pandemic of 1918. Available at: http://virus.stanford.edu/uda/.
5. Sherman IW. Twelve diseases that changed our world. New York: ASM Press; 2007.

Silence That Goes Beyond Words

Those who have the strength and love to sit with a dying patient in the silence that goes beyond words will know that this moment is neither frightening nor painful, but a peaceful cessation of the functioning of the body. Watching a peaceful death of a human being reminds us of a falling star; one of a million lights in a vast sky that flares up for a brief moment only to disappear into the endless night forever. To be a therapist to a dying patient makes us aware of the uniqueness of each individual in this vast sea of humanity. It makes us aware of our finiteness, our limited lifespan. Few of us live beyond our three score and 10 years and yet in that brief time most of us create and live a unique biography and weave ourselves into the fabric of human history.

Psychiatrist and author Elizabeth Kübler-Ross, MD (1926–2004) [1]

The words above are the last paragraph in the Elizabeth Kübler-Ross book, *On Death and Dying*, published in 1969. In her groundbreaking work, the author elucidates the five stages of grief. I first read this book as a young physician, and even

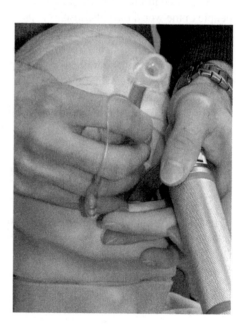

Fig. 4.15 Endotracheal intubation: an all-too-common final futile procedure. This file is licensed under the Creative Commons Attribution 3.0 Unported license. http://commons. wikimedia.org/wiki/ File:Intubation_endotra-cheal_tube_laryngoscope.jpg

today can recite the stages described as "coping mechanisms at the time of a terminal illness:" (1) denial and isolation, (2) anger, (3) bargaining, (4) depression, and (5) acceptance [1]. That I have kept my original copy of the book through a number of geographic moves, including two across America, is testimony to the value I have placed on it. What, if anything, has changed since 1969?

Today I wonder if we are seeing as many peaceful deaths as we should. In the late 1960s, when the Kübler-Ross book was first published, I was a small-town solo family physician. Through house calls and close coordination with family members, many peaceful home deaths occurred. Also, at about this time we saw the beginning of the hospice movement in England (described next) and later in America, including "home hospice," supporting those who chose to die where they had lived.

In a 1989 survey of 90 patients facing death, 48 (53 %) wished to die at home and only 13 (14 %) preferred to die in hospital [2]. Yet what happens today is typically the obverse of what patients and their families would want. Our health care system, in which high-tech medicine is the norm and home care a luxury, finds ways to channel dying persons into sterile, expensive hospital beds. One report estimates that end of life care consumes one quarter of America's health care budget [3]. Few would regard this as a wise use of our nation's resources. In the hospital setting, the experience of dying can become, again in the words of Kübler-Ross, "lonely, mechanical, and dehumanized." The patient's final moments may be a frenzy of eager chest-thumping, fumbling attempts to insert access lines, and inexperienced interns seizing an opportunity to practice intubation—just the opposite of peaceful (see Fig. 4.15).

In 2004 Kübler-Ross died at age 78 following a series of debilitating strokes. Her death notice in the New York Times reports that in her 1997 autobiography, she penned: "I always say that death can be one of the greatest experiences ever" [4]. She seemed to have reached the stage of "acceptance." Kübler-Ross died at home; I hope her final hours were peaceful.

1. Kübler-Ross E. On death and dying. New York: Macmillan;1969, page 276.
2. Dunlop RJ et al. Preferred versus actual place of death: a hospital palliative care support team experience. Palliat Med. 1989;3:197.
3. Becker G et al. Terminal care and the value of life near its end. NBER Working Paper No. 15649 January 2010.
4. Death notice: Elizabeth Kübler-Ross. New York Times. Available at: http://www.nytimes.com/2004/08/26/us/elisabeth-kubler-ross-78-dies-psychiatrist-revolutionized-care-terminally-ill.html?pagewanted=all&src=pm.

Accepting the Inevitable

> The old acceptance of destiny has gone, and a new sense of outrage that modern advances cannot finally halt the inevitable make care of the dying and their families demanding and often difficult, but perhaps all the more rewarding.
>
> English physician, nurse, and social worker Dame Cecily Saunders (1918–2005) [1]

In 1980, writing in the *New England Journal of Medicine*, Fries concluded: "Present data allow calculation of the ideal average life span, approximately

Fig. 4.16 St. Christopher's Hospice. The copyright on this image is owned by Stephen Craven and is licensed for reuse under the Creative Commons Attribution-ShareAlike 2.0 license. http://commons.wikimedia.org/wiki/File:St._Christopher%27s_Hospice.jpg

85 years." He went on to predict, citing modern medical advances and the fact that the maximum life span has not increased, "that the number of very old persons will not increase, that the average period of diminished physical vigor will not increase, that chronic disease will occupy a smaller proportion of the typical life span, and that the need for medical care in later life will decrease" [2]. According to Fries, we have eliminated 80 % of years of life previously lost to premature death, allowing survival curves to take on a "rectangular" form, describing living in good health until age in the mid-1980s followed by a steep decline to death [2].

If our public health measures and wondrous technology can reduce the proportion of the life span occupied by chronic diseases, then why can't we halt the inevitable journey to death? Saunders describes the new sense of resentment over medicine's seeming failure to conquer the last frontier—death. As I read her description of "outrage," I think of Kübler-Ross' five stages of coping (denial, anger, bargaining, depression, and acceptance), especially the first two: *denial* and *anger* [3].

But eventually patients and families pass through the earlier stages of grief to reach *acceptance*. Part of this process is often influenced by the progression of disease—as the patient moves from independence, to needing help, to dependence, to the terminal stages of life. Somewhere near the end of this process, hospice care often becomes desirable. And for the availability of hospice, we can thank Dame Cecily Saunders, who pioneered the modern hospice movement with the opening in 1967 of the St. Christopher's Hospice in South London, England. From there the hospice movement has spread across the UK, USA, and many other countries (see Fig. 4.16).

Hospice care offers the terminal patient emotional and spiritual care, and especially the alleviation of symptoms, such as pain, dyspnea, and nausea. As Saunders and those that followed her discovered, once the patient and family make peace with the futility of "curative" treatment for terminal disease, and begin to focus on orchestrating the best death possible, the doctors and nurses find their work highly rewarding.

1. Saunders C. Foreword. In: Doyle D et al., eds. Oxford textbook of palliative care. Oxford: Oxford University Press; 1993.
2. Fries JF. Aging, natural death, and the compression of mortality. N Engl J Med. 1980;303;130.
3. Kübler-Ross E. On death and dying. New York: Macmillan; 1969, page 276.

Chapter 5
Disease Detection and Diagnosis

Just as I began work on this chapter I happened to read a "Perspective" article by Croskerry in the *New England Journal of Medicine*. The essay begins: "The two major products of clinical decision making are diagnoses and treatment plans. If the first is correct, the second has a greater chance of being correct too. Surprisingly, we don't make correct diagnoses as often as we think: the diagnostic failure rate is estimated to be 10–15 %." The author goes on to assert that the risk of diagnostic error is greatest among those specialists most likely to encounter patients with undifferentiated problems—those in emergency medicine, family medicine, and (general) internal medicine [1]. The reason for the higher risk of misdiagnosis among these specialties is not that the practitioners lack diagnostic acumen, but rather because they are on the front lines, engaged in "first-contact" encounters with many patients presenting vague symptoms and early clinical signs, and whose disease causes have yet to have labels attached.

Diagnosis—the elucidation of the cause of heretofore unexplained symptoms, signs, and laboratory/imaging findings—must be one of the highest forms of human cognitive function. Once I have the diagnosis, I can readily look up treatment of the disease. As to prevention of specific diseases, there are myriad recommendations and guidelines published. But diagnosis remains a cerebral exercise of "connecting the dots" from chief complaint to confirmed disease cause.

Currently there is fascination with computer-assisted diagnosis, an approach in evolution since the 1970s. An early commercial program, based on work at the University of Pittsburgh, was marketed as *Quick Medical Reference*. A currently popular differential diagnosis tool called *Isabel* (named for the daughter of the creator) was described in a 2012 article in the New York Times [2].

But still, computer-assisted diagnosis seems unlikely to replace the function of the educated human brain. On a technical basis, computer-assisted programs

© Springer Science+Business Media New York 2015
R.B. Taylor, *On the Shoulders of Medicine's Giants*,
DOI 10.1007/978-1-4939-1335-0_5

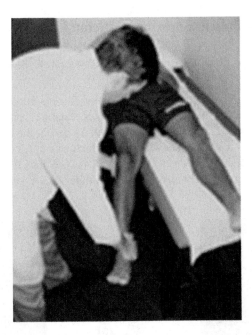

Fig. 5.1 Physical examination of a right knee—notice distal hand applying valgus force to foot. http://www.springerimages.com/Images/MedicineAndPublicHealth/1-10.1007_s12178-007-9016-x-2

integrate current discoveries and published data, which means that—like medical reference books—they are always a little out of date. Also, will a computer ever be able to perform a useful examination of, for instance, the knee? (see Fig. 5.1.)

Then there is the computer vs. brain debate. Computers can best even elite chess players. But chess and clinical diagnosis differ. The computer cannot recognize the patient's unspoken responses to questions, the flashes of emotion, the occasional purposeful evasion, and the other subtle clinical cues that are often the keys to the correct diagnosis. Diagnosis is more than the movement of rooks and bishops on a board.

In this chapter, we see what some of medicine's giants—from the fifth century BCE to much more recent times—have had to say about diagnosis.

1. Croskerry P. From mindless to mindful practice—cognitive bias and clinical decision making. N Engl J Med. 2013;368:2445.
2. Hafner K. Could a computer outthink this doctor? The New York Times. Tuesday Dec. 2, 2012, page D1.

The Doctor as Patient and Diagnostician

By observing myself I know about others and their diseases are revealed to me ....
Chinese sovereign Huang Ti (The Yellow Emperor) (ca. twenty-sixth century BCE) [1]

In 1973, I wrote a health care guide for senior citizens titled *Feeling Alive after 65* [2]. My book described hypertension, strokes, cataracts, prostate enlargement, and more. At that time I was all of 37 years old. As part of the book's promotion, I appeared on various television shows, one of which was *AM New York*. I recall that during that appearance there was a time for call-in questions from the TV viewing audience. One caller (undoubtedly a good deal older than I was at the time) asked

余嘗上於清冷之台，
中階而顧，匍匐而前，
則惑。

Fig. 5.2 Huang Ti (the Yellow Emperor). http://www.springerimages.com/Images/MedicineAnd PublicHealth/1-10.1007_s00415-012-6523-5-0

somewhat testily, "You're just a young kid. How could you know anything about these diseases you are talking about that affect us older people? You've never had any of them."

Huang Ti, if he existed at all and if legend is correct, lived approximately 100 years, long enough to experience a host of maladies associated with aging. We don't know exactly when the Yellow Emperor's book [1] was written and, like the *Corpus Hippocraticus*, it may well have been compiled in some later era. But if we are to assume the veracity of the author's long life, and his opportunity to "observe myself" through various diseases, then we can understand how he believes this experience allows "their diseases [to be] revealed to me" (see Fig. 5.2).

When I was a senior in medical school, my wife and I were parents of an infant daughter; this was very fortunate because, in the pediatric clinic, I was counseling young mothers on routine child care: "How do I recognize colic?" "What does it mean if my baby seems to spit up after every feeding?" "How high a temperature elevation merits a trip to the emergency room?" In answering these basic diagnosis questions, I relied heavily what I was learning by personal experience. I wondered how my childless classmates could deal with these questions.

During my practice years, I told kidney stone sufferers: "Yes, passing a kidney stone can be more painful than childbirth." Of course I had experienced neither. I treated patients with asthma, angina, depression, heart failure, and cancer—having never suffered any of these ailments.

I recommend Dr. Edward Rosenbaum's book *A Taste of My Own Medicine*, chronicling what happens when a successful physician develops cancer of the larynx. He describes a delayed diagnosis, physician indifference to his misery, and the isolation that can be comprehended only by those who suffer life-threatening illness [3]. Or watch the 1991 movie *The Doctor*, loosely based on the Rosenbaum novel, but with the same strong message. Reading a book or seeing a movie is a sorry substitute for having disease oneself, but until the young physician has logged the years that bring personal experience with various maladies, the vicarious experience of ailments may help improve diagnosis—and management.

1. Huang Ti. *Nei ching su wen*: The yellow emperor's classic of internal medicine. Book 2, sect. 2. Veith I, trans. Berkley: California Press; 1966.
2. Taylor RB. Feeling alive after 65: the complete medical guide for senior citizens and their families. New Rochelle, NY: Arlington House; 1973.
3. Rosenbaum EE. A taste of my own medicine: when the doctor is the patient. New York: Random House; 1988.

The Role of Patience in Diagnosis

> Now they say when I come to a patient, I know not immediately what ails him, but I need time to find out. It is true. That they judge immediately is the fault of foolishness; for in the end the first judgment is false and from day to day they know the longer, the less, what it is, and make liars of themselves. Whereas I desire to approach from day to day, the longer, the closer to the truth.
>
> Renaissance physician Paracelsus (1493–1541) [1]

In Chap. 1, I told that Philippus Aureolus Theophrastus Bombastus von Hohenheim assumed the name Paracelsus to indicate that he was "equal to or greater

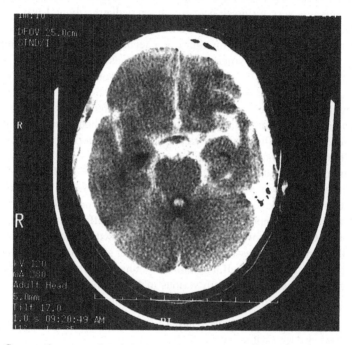

Fig. 5.3 Computed tomography of the brain demonstrates a subarachnoid clot in the inter-hemispheric fissure, suprasellar and ambient cisterns, and sylvian stem from a middle cerebral artery aneurysm rupture. http://www.springerimages.com/Images/MedicineAndPublicHealth/2-ACA0101-09-010

than" the Roman encyclopedist Aulus Cornelius Celsus (25 BCE–50 CE). Paracelsus' hubris did not end with his name change, and despite work that has earned him the title "Father of Toxicology," he attracted a legion of detractors, prompting him to write his book, best known today as *The Seven Defenses of Paracelsus: (Against Those Who Would Seek to Destroy Me)* [1].

Despite his reputed arrogance, thoughts of Paracelsus quoted above strike me as quite reasoned, even humble. He makes no claim to being able to diagnose disease in an instant, and instead calls for allowing a gradual approach to detect the cause of a patient's symptoms and signs.

Deciding immediately what ails a patient is called a "curbstone diagnosis"—a danger for any health professional who ventures a premature medical opinion. Here I am not writing about two physicians discussing a case over coffee—the helpful, typically informal and unrecorded, consultation we physicians often seek from our colleagues. I am writing about the "quickie diagnosis" that may occur in a chance encounter in the hallway or even in a social setting. There is a dearth of papers about this important topic, but I did find an amusing article in the *Toledo Blade*, titled "Curbstone Diagnosis Costs Nothing, Usually Worth Just as Much." In this article, the physician author tells of a physician who made a curbstone diagnosis of hip arthritis in a golf club locker room; the golfing buddy was found a week later to have pancreatic cancer metastatic to the hip. "My colleague was known thereafter as the physician who did not know the difference between cancer and arthritis" [2].

Gray et al. have studied various aspects of continuity care. They conclude: "Continuity of care is associated with better diagnosis" [3]. Experienced generalists, the experts in continuity care, become familiar with the concept of the "continuously evolving and tentative diagnosis," meaning that the disease you see at today's visit will never be quite the same when the patient returns next month. A nagging cough may eventually reveal itself to be asthma, or have subsided completely. A headache, generally an easy diagnosis, occasionally turns out to be caused by an intracranial bleed. Sometimes diagnostic detection requires applying a "tincture of time" (see Fig. 5.3).

1. Paracelsus. Seven defenses,1535. Sequim WA: Holmes Publ Group; 2001.
2. Van Dellen TR. Curbstone diagnosis costs nothing, usually worth just as much. Toledo Blade; May 20, 1957, page 9.
3. Gray DP et al. Towards a theory of continuity of care. J Roy Soc Med. 2003;96:160.

On Innovation in Diagnostic Methods

In 1816 I was consulted by a young woman laboring under general symptoms of a diseased heart, and in whose case percussion and the application of the hand were of little avail on account of the great degree of fatness .... (Direct auscultation with the ear to the chest wall) being rendered inadmissible by the age and sex of the patient, I happened to recollect a simple and well-known fact in acoustics, ... the great distinctness with which we hear the scratch of a pin at one end of a piece of wood on applying our ear to the other. Immediately, on this suggestion, I rolled a quire of paper into a kind of cylinder and applied one end of it to the region of the heart and the other to my ear, and was not a little surprised and pleased to find that I could thereby perceive the action of the heart in a manner much more clear and distinct than I had ever been able to do by the immediate application of the ear.

French physician René Laennec (1781–1826) [1]

Medical students learn the four classic maneuvers in physical examination of the patient: *Inspection, Palpation, Percussion,* and *Auscultation.* Inspection, as critical observation of the patient, was used by the ancient Greeks and championed by Hippocrates. I could not find any historical beginning date of palpation as a diagnostic method but, as an amateur etymologist, I note that the word "palpate," from the Latin *palpatus*, meaning to stroke or touch, came into use about 1840–1850 [2].

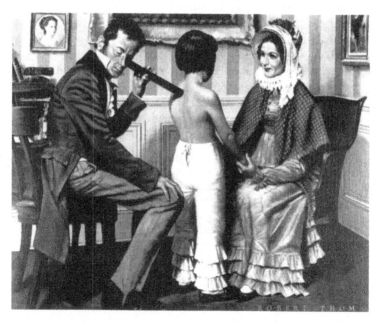

Fig. 5.4 Indirect auscultation of body sounds with a Laennec stethoscope. http://www.springer images.com/Images/Engineering/1-10.1007_978-3-642-24843-6_1-10

Fig. 5.5 Collins sign demonstrated to Professor Collins by patient with a hand behind the back and the thumb pointing upwards

Austrian physician Leopold Auenbrugger (1722–1809) is credited with introducing percussion to the physician's repertoire. He examined the patient's chest using a technique similar to how he tested for the content of wine barrels in the cellar of his father's hotel. Laennec, aware of Auenbrugger's work, popularized auscultation as the fourth member of the physician's diagnostic quartet, earning himself a niche in the pantheon of medical giants (see Fig. 5.4).

Over the years many physicians have developed useful diagnostic maneuvers: Think of the Romberg test (German physician Moritz Romberg, 1795–1873); Chvostek sign (Czecho-Austrian physician František Chvostek, 1835–1884); Babinski reflex (French neurologist Joseph Babinski, 1857–1932); and Tinel sign (French neurologist Jules Tinel, 1879–1952).

New physical examination methods are still being devised. Consider the Dix-Hallpike test for benign paroxysmal positional vertigo described in 1952 [3]. And more recently the Collins sign, described by Dr. Paddy Collins in Ireland, in which patients with acute cholelithiasis often describe the discomfort by placing the hand behind the back with the thumb pointing to the tip of the scapula [4] (see Fig. 5.5). Perhaps some reader of this book will add a new technique to our toolbox of diagnostic methods.

1. Laennec R. *De l'auscultation mediate*. Paris: Brosson and Chaude; 1819.
2. Random House Dictionary. New York: Random House; 2013.
3. Dix MR, Hallpike CS. The pathology, symptomatology, and diagnosis of certain common disorders of the vestibular system. Proc. R. Soc. Med. 1952;45:341.
4. Gilani SN. Collins sign: validation of a clinical sign in cholelithiasis. Irish Med J. 2009;178:397.

What We Learn by Just Looking

The trouble with doctors is not that they don't know enough, but that they don't see enough.

Irish physician Sir Dominic Corrigan (1802–1888) [1]

Sir Dominic Corrigan's career progressed from private practice in Dublin to the roles of president of the Irish College of Physicians and physician-in-ordinary to the Queen in Ireland. His pioneering work on the manifestations of heart disease included the description of a carotid pulse that is bounding and forceful, then suddenly collapsing, seen with aortic regurgitation, and known today as the Corrigan pulse [2].

When I read the Corrigan quotation above, I think of the homespun wisdom of American baseball player, Yogi Berra, who once remarked, "You can observe a lot by just watching." Here is an example: The patient was a 32-year-old woman visiting her

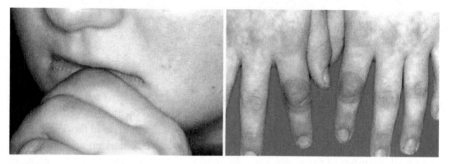

Fig. 5.6 Russell sign: calluses of the knuckles seen in some patients with bulimia nervosa. http://www.springerimages.com/Images/MedicineAndPublicHealth/1-10.1007_978-1-4614-1111-6_16-0

physician for her annual Pap smear. As part of the routine examination, the physician noted that the patient, who had always been of normal body habitus, had lost 18 pounds in weight over the past year. The patient offered no explanation. A routine physical examination revealed only evidence of recent weight loss. The doctor considered the usual suspects in cases of unintentional weight loss: cancer, infections such as acquired immunodeficiency syndrome (AIDS) or tuberculosis, depression, hyperthyroidism, and so forth. As he sat back to plan the first round of laboratory tests and imaging to determine the diagnosis, he happened to notice her fingers.

On the dorsum of the knuckles of the left hand were thick calluses. The doctor recalled reading about such a finding: The patient had the Russell sign, named for British psychiatrist Gerald Russell, who published his description of bulimia nervosa in 1979 [3]. The calluses of Russell sign classically occur as a patient with an eating disorder, typically bulimia nervosa, uses the fingers to induce vomiting, thereby causing the maxillary incisors to repeatedly abrade the skin of the knuckles (see Fig. 5.6).

Although I am not a proponent of premature diagnoses, the observant physician can often suspect the disease present simply by observation of the patient's face. Consider the masklike facies of parkinsonism, the "moon" face of Cushing syndrome, the open-mouth countenance of children with adenoidal hypertrophy, the chipmunk appearance of beta-thalassemia major or bilateral parotid swelling, the prognathism and frontal bossing of acromegaly, and the fabled leonine facies of advanced leprosy [4].

All this and more can be seen by just looking.

1. Corrigan DJ. Lectures on the nature and treatment of fever. Dublin: Fannin and Co; 1853.
2. Corrigan DJ. On permanent patency of the mouth of the aorta, or inadequacy of the aortic valves. Edinburgh Med Surg J. 1832;37:225.
3. Russell G. Bulimia nervosa: an ominous variant of anorexia nervosa. Psychological Med. 1979;9:429.
4. Taylor RB. Diagnostic principles and applications: New York: Springer; 2013.

In Search of the Whole-Patient Diagnosis

There used to be a French saying that "French physicians treat the disease, English the patient." So far as this is true it is to the honor of the English, for to treat a sick man rightly requires the diagnosis not only of the disease but of all manner and degrees in which its supposed essential characters are modified by his personal qualities, by the mingled inheritances that converge in him, by the changes wrought in him by the conditions of his past life, and by many things besides.

English surgeon Sir James Paget (1814–1899) [1]

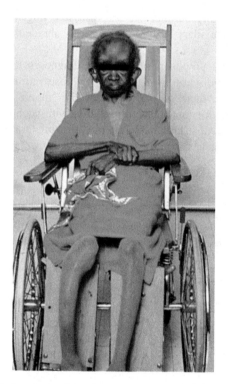

Fig. 5.7 This patient displays many of the classic features of Paget disease, including severe bowing of the right lower leg, frontal bossing, and deafness as manifested by her hearing aid. http://www.springerimages.com/Images/MedicineAndPublicHealth/1-10.1007_978-1-4614-1111-6_1-3

Remembered today as the namesake for several diseases, notably Paget disease of bone, Paget was a scientist as well as a surgeon. Paget's thinking, as reflected in the quotation above, was a prelude to some changes we see in medicine today.

A few decades later, Sir William Osler observed: "It is much more important to know what sort of a patient has a disease than to know what kind of a disease a patient has" [2]. A diabetic patient with the means to afford proper medication and follow a healthy diet is quite different from a diabetic patient who lives from paycheck to paycheck and may have to choose between medication and food. Such differences are actually part of the whole-patient diagnosis, because they profoundly affect disease management. In the picture of the patient with Paget disease (see Fig. 5.7), the diagnosis would be greatly enriched by identification of her personal qualities, mingled inheritances, and the conditions of home and her past life.

The approach used is often termed *biopsychosocial*, a model pioneered by George Engel, who wrote in 1978 of the average physician of his day that, despite impressive technical skills, "when it comes to dealing with the human side of illness and patient care displays little more than the native ability and personal qualities with which he entered medical school" [3]. Few of us who graduated from medical school a few decades ago or more had any instruction in how to explore the patient's personal qualities and integrate them into diagnosis and treatment. The patient was little more than the passive host of some interesting pathologic abnormalities.

Today's medical school admissions are paying more attention to the ability of applicants to relate to others and the world about them [4]. The curriculum recognizes the patient as a person, and strives to make young doctors better interviewers than their predecessors. There are currently medical school workshops aimed to help improve communication skills, observed interviews of "standardized patients," and even clerkships emphasizing psychosomatic medicine [5]. We may yet see the end of the stereotypic physicians that, according to Engel, patients described as "insensitive, callous, neglectful, arrogant and mechanical in their approaches" [3].

1. Paget J. Address to the Abernethian Society, 1885. Quoted by Sir James Patterson Ross. Bartholomew's Hosp J. 1950;54:50.
2. Osler W. Osler publishes the Principles and Practice of Medicine, 1892. Available at: http://www.thenewmedicine.org/timeline/doctor_patient_book.
3. Engel GL. The biopsychosocial model and the education of health professionals. Ann New York Acad Sciences. 1978;310:169.
4. Muller D. Reforming medical education—out with the old, in with the new. N Engl J Med. 2013;368:1567.
5. Bourgeois JA et al. Reflections on psychosomatic medicine as a third-year medical student clerkship: an integrated experience that demonstrates the biopsychosocial model. Acad Psych. 2012;36:240.

The Successful Diagnostician

> The precise and intelligent recognition and appreciation of minor differences is the real essential factor in all successful medical diagnosis. Eyes and ears which can see and hear, memory to record at once and to recall at pleasure the impressions of the senses, and an imagination capable of weaving a theory or piecing together a broken chain or unraveling a tangled clue, such are the implements of his trade to a successful diagnostician.
>
> British physician and educator Joseph Bell (1837–1911) [1]

How good are you at weaving a theory or unraveling a tangled clue? Here is a quiz to test your abilities as a diagnostician. Consider the implications of the following clinical findings.

1. What disease should you consider in a child with nasal polyps?

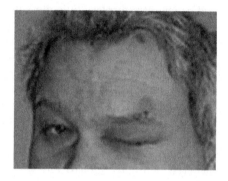

Fig. 5.8 Herpes zoster ophthalmicus (Public domain). http://commons. wikimedia.org/wiki/ File:Herpes_Zoster_im_ Augenbereich.JPG

2. What disorder is likely when a man tells you that his fifth finger repeatedly "catches" when he puts his hand in his trouser side pocket?
3. When a patient has pruritic skin lesions in the pattern of the wake left by a moving ship, what disease should you consider?
4. What cause of back pain often improves when walking up steps?
5. What severe condition may follow painful blisters on the tip of the nose?

Answers [2]:

1. Cystic fibrosis, the most common lethal genetic disease of Caucasian persons.
2. Ulnar nerve palsy—the Wartenberg sign.
3. Scabies—the "wake" sign.
4. Spinal stenosis, a disorder in which forward flexion relieves nerve pressure.
5. Herpes zoster ophthalmicus—the Hutchison sign (see Fig. 5.8).

We know these facts because of the observations of astute clinical diagnosticians who have gone before us.

The point to be made here is this: Diagnosis, at its core, is a hands-on, cerebral, personal endeavor. To be a successful diagnostician, you need to be a skilled physical examiner, an able clinician who is up to date on current knowledge yet acquainted with traditional diagnostic methods, a possessor of a good memory who can recall the implications of seldom encountered findings, and an empathic listener who can patiently allow the patient to tell you—or perhaps show you—the diagnosis. Bell incidentally was the inspiration for Arthur Conan Doyle's fictional character, Sherlock Holmes, discussed on the following page.

1. Bell J. The adventures of Sherlock Holmes: a review. Bookman; 1892, page 79.
2. Taylor RB. Diagnostic principles and applications. New York: Springer; 2013.

Seeing What Others Overlook

"Never mind," said Holmes, laughing: "it is my business to know things. Perhaps I have trained myself to see what others overlook. If not, why would you come to consult me?"

Scottish physician and author Arthur Conan Doyle (1859–1930) [1]

The most difficult diagnoses often rest on the recognition of trifles, the small things that others may see but don't recognize as telltale signs of disease. This is how Sherlock Holmes solved the difficult cases. As you can deduce from the quote above, Holmes was not a modest man, but he was good at his job.

It may come as no surprise that Sherlock Holmes was the brainchild of a physician, Sir Arthur Conan Doyle. Not long after graduating from the University of Edinburgh, he (Doyle, not Holmes) established a private practice in the resort town of Seaside in the county of Hampshire in England. As is often the case with a new medical practice, Dr. Doyle had some long hours between patients, and he filled them by writing (see Fig. 5.9).

Fig. 5.9 Sir Arthur Conan Doyle. http://www.springerimages.com/Images/MedicineAndPublicHealth/1-10.1007_978-1-4614-0170-4_1-27

In an interview in 1892, Doyle tells how he happened to envision "consulting detective" Sherlock Holmes, based on his experience with Dr. Joseph Bell: "Oh! But, if you please, he (Sherlock Holmes) is not evolved out of anyone's inner consciousness. Sherlock Holmes is the literary embodiment, if I may so express it, of my memory of a professor of medicine at Edinburgh University, who would sit in the patients' waiting room with a face like a red Indian and diagnose the people as they came in, even before they opened their mouths. He would tell them their symptoms, he would give them details of their lives, and he would hardly ever make a mistake. 'Gentlemen,' he would say to us students standing around, 'I am not quite sure whether this man is a cork-cutter or slater. I observe as slight callus, or hardening on one side of his fourth finger and a little hardening on the outside of his thumb, and that is a sure sign he is either one or the other.' His faculty of deduction was at times highly dramatic. 'Ah!' he would say to another man, 'You are a soldier, a noncommissioned officer, and you have served in Bermuda. Now how did I know that, gentlemen? He came into the room without taking his hat off, as he would going into an orderly room. He was a soldier. A slight authoritative air, combined with his age, showed he was an N.C.O. (noncommissioned officer). A slight rash on the forehead tells me he was in Bermuda, and subject to a certain rash known only there.' So I got the idea for Sherlock Holmes" [2].

Perhaps Doyle's recollection of Bell's abilities was influenced a little by the awe that can attend the student-professor relationship, and I am just a little skeptical about a forehead rash peculiar to Bermuda. But there can be no question that the best diagnosticians are those who recognize small bits of evidence. Here are just a few: urine that turns dark upon exposure to sunlight that can occur with porphyria; bulky and hard-to-flush stool in persons with celiac disease; the slowness to walk or the later use of hands to "walk up the legs" when rising to a standing position seen in children with muscular dystrophy; the bitemporal hemianopsia that can be the tip-off to the presence of a pituitary tumor; the hyperpigmentation of the skin seen in 70 % of patients with hemochromatosis; and the loud snoring and daytime sleepiness that can suggest the presence of obstructive sleep apnea [3].

1. Doyle AC. A case of money (1891). In: The adventures of Sherlock Holmes. Herfordshire, England: Wordsworth Editors; 1992.
2. Blathwayt R. A talk with Dr. Conan Doyle. The Bookman. 1892.2:50.
3. Taylor RB. Diagnostic principles and application. New York: Springer; 2013, page 157.

Attending to Clues Wherever Found

To throw open the mind's door and allow diseases to enter into consideration each time we are called to a bedside is foolish in the attempt, and impossible in the performance. Each case should lead us to arrange before the mind's eye a selected group of reasonably probable causes for the symptoms complained of and for the signs discovered. What we select should depend on the clues furnished us by the patient himself or by the results of our own examination.

American physician Richard C. Cabot (1868–1939) [1]

Dr. Richard C. Cabot, author of the 1915 book *Differential Diagnosis*, provided care for the medically underserved of Boston until 1919, when he became chairman of the Harvard University Department of Social Ethics and founded the

Fig. 5.10 Richard C. Cabot, founder of the MGH social service department, with Ida Cannon, its longtime director. http://www.springerimages.com/Images/MedicineAndPublicHealth/1-10.1007_s12682-011-0091-9-1

Massachusetts General Hospital (MGH) social service department [2] (see Fig. 5.10). Is there something odd—a cognitive disconnect—in encountering a physician who writes one of our early reference works on differential diagnosis and then who decides to devote his career to medical ethics and social work? Perhaps not.

First of all, consider Cabot's admonition not to "open the mind's door and allow diseases to enter into consideration" with each patient encounter. In the office, as I consider the possible cause of a patient's undifferentiated symptoms or signs—such as recurring fever, fatigue and lethargy, or loss of appetite—I find that trying to deal with more than four or five entities in my mind becomes difficult. I need to quickly eliminate the improbable causes and focus on the likely etiologies. The patient with continuing anorexia may have depression, an eating disorder, a malignancy, or per-haps an infection such as viral hepatitis or AIDS. I don't consider chronic kidney disease, thalassemia, Addison disease, hypervitaminosis D, or zinc deficiency—unless there is some special reason to do so.

In the USA today, tuberculosis is unlikely to merit inclusion in the "short list" of differential diagnosis possibilities for the patient with anorexia. That is, unless the patient is a recent immigrant from a country, such as Korea, where there is a high incidence of tuberculosis [3]. Or unless the patient is a young person with impaired appetite who lives with a grandparent with unidentified active tuberculosis.

Here lie some of the other intersections between differential diagnosis and the social sciences. Maternal stress is a risk factor for preterm obstetrical delivery and low birth weight [4]. So is low socioeconomic status, which must carry some degree of stress. The patient with recurrent pelvic pain, if questioned carefully, will often describe a history of childhood sexual abuse. In a patient with chronic complaints of abdominal pain and bloating, learning of recent personal, professional, or financial problems, such as job loss, divorce, or death, may help support a diagnosis of irri-table bowel syndrome [5].

And so when faced with a diagnostic challenge, try early to limit the possibilities being considered to "a selected group of reasonably probable causes," and don't forget that important diagnostic clues may be those that are sometimes found in the past history and current everyday life of the patient.

1. Cabot R. Differential diagnosis. Philadelphia: Saunders; 1915.
2. Bull W. Massachusetts General Hospital: Noblesse oblige, private practice, and the power of research. J Med and Person. 2011;9:80.
3. Kim HJ. Current status of tuberculosis in Korea. Korean J Med. 2012;82:257.
4. Hobel CJ et al. Psychosocial stress and pregnancy outcome. Clin Obstet Gyn. 2008;51:333.
5. Holten KB. Diagnosing the patient with abdominal pain and altered bowel habits: is it irritable bowel syndrome? Am Fam Phys. 2003;67:2157.

The Medical History as a Work of Art

Having acquired a patient, the first thing to do is obtain a history of his ailment. The secur-
ing of an adequate one is a work of art. It requires a knowledge of disease and of human
nature. It is hard work and is time-consuming but it is necessary, because in many cases it
is the most important factor in the whole procedure. A good history may even anticipate
what the microscopic slide will show.

American physician Arthur E. Hertzler (1870–1946) [1]

Shall we include Dr. Hertzler among the pantheon of medicine's all-time greats?
We think of him as the country doctor who wrote a book describing his days serving
as physician to the residents of Halstead, Kansas. So why include him here? I can
think of three reasons.

First of all, he was much more of an academic scholar than most realize. Writing
of Hertzler in 1944, Dr. Logan Clendening states: "He posed as a common-sense
horse and buggy doctor. He had common sense all right, but those of us who knew
him disregarded his pose because, to our knowledge, he was one of the best-
informed, most brilliant, and learned surgeons of our country" [2]. In a PubMed
search, I found five scholarly articles that he authored and published in peer-
reviewed surgery journals. This does not count the 20 medical books he authored.

Secondly, Hertzler influenced a generation of Americans with his book *Horse
and Buggy Doctor*, full of homespun wisdom about medical care as service to
people.

The book, written for the general audience and not for physicians, rose to number
five on the nonfiction bestseller list (ahead of Dale Carnegie's *How to Win Friends
and Influence People*) and enjoyed an astounding 45 printings [3, 4]. Today, copies
of Hertzler's books are often found on the shelves of used book stores. Would it be
unreasonable to suggest that the nostalgia for the "Old Doc" generated by *Horse
and Buggy Doctor* helped set the stage for the rise of family medicine as a specialty
in the 1960s and 1970s? I think that the article by Coulehan suggests this [4].

Third, and with the above facts as background, I selected Hertzler to be in this
book because I liked his typically down-to-earth comments about the medical history.

A good medical history is, indeed, a work of art. And, as with any quality work, it takes time. One of the many obstacles to obtaining a superior medical history is the time pressure often felt by physicians. This sense of time urgency can lead to two actions, neither likely to enhance the final product. The first is delegation of history-taking to someone else, so that by the time the doctor enters the room the chief complaint, history of present illness, and all the rest are already entered on the computer screen. The physician need only jump in mid-stream: "I see you are here about your stomach pains." But by using this style, the physician has missed the facial expressions and other nonverbal cues that were not seen at all as an assistant transcribed answers to scripted questions.

The second enemy of a good medical history is interruption by the physician. Groopman holds that doctors typically interrupt the patient's narrative after 18 s [5]. Here is what family physician educator G. Gayle Stephens says about physician interruptions: "There is a golden moment at the beginning of each visit when the patient's priorities and uncoached words are full of possibilities for disclosing his or her illness. The moment you ask a leading question, the possibilities diminish" [6]. This tendency to "jump in" and ask closed-ended questions clearly speeds the interview, but often at the expense of diagnostic accuracy.

In fact, the most useful historical clues often come with the opposite of interruption: silence. The experienced interviewer learns the value of silence. To fill the uncomfortable void, the interviewee—the patient—will often take the conversation down a totally unexpected path. And this path may be just the highway that will lead you to the diagnosis.

1. Hertzler AE. The horse and buggy doctor. New York: Harper; 1938.
2. Arthur E. Hertzler, MD. Kansas City Times, August 1, 1970. Available at: http://www.kchistory. org/cdm4/item_viewer.php?CISOROOT=/Mrs&CISOPTR=1144.
3. Jarrott SD. Horse and buggy doctor. Private Pract. May, 1971; page 24.
4. Coulehan J. Is there a Doctor Hertzler in the house? The Pharos. Autumn, 2012; page 18.
5. Groopman J. How doctors think. New York: Houghton Mifflin; 2007.
6. Stephens GG. In: A little book of doctors' rules II. Meador CK, ed. Philadelphia: Hanley & Belfus; 1999, Rule number 89.

Common Things Occur Most Commonly

If you hear hoof beats behind you and you turn around, expect to see horses and not zebras.

American physician/researcher Theodore E. Woodward (1914–2005) [1]

Dr. Woodward accomplished a great deal in his life, in addition to giving young doctors one of their most colorful and memorable aphorisms. He was an infectious disease specialist who held the post of Chairman of the Department of Internal Medicine at his alma mater, the University of Maryland School of Medicine in Baltimore, Maryland, from 1954 to 1981. His work on finding effective treatments for typhus and typhoid fever during World War II earned him a Nobel Prize nomination.

In fact, Woodward probably never said exactly what I wrote above, and I cannot find that he ever wrote the adage in any published work. I have found two sources crediting Woodward with the "zebra" saying, one online citation [2] and one in print [1]. Richard Colgan, a professor at the Woodward's medical school and who personally spent time with him, quotes him as follows: "If you are walking down Green Street (the street in front of University Hospital in Baltimore) and you hear hoof beats behind you, don't look back expecting to see a zebra. Expect a horse" [1].

Fig. 5.11 Myasthenia gravis. Photo credit: James Heilman. This file is licensed under the Creative Commons Attribution-Share Alike 3.0 Unported license. http://commons.wikimedia.org/wiki/File:DiplopiaMG1.jpg

What might be a *zebra diagnosis*? As an example, I think that Dercum disease, with multiple painful lipomas, would qualify as zebra diagnosis.

The zebra metaphor is firmly rooted in our medical lexicon as representing an unexpected and sometimes improbable diagnosis, especially when a more common cause is most likely. The term is often used in medical lectures without further elucidation of the metaphor. There is even a publication on this topic [3].

But zebras are not unicorns. We must keep in mind that zebras exist in the world, and zebra diagnoses occur in medicine. Sooner or later a menopausal patient with flushing will have a carcinoid tumor and a man with shortness of breath on exertion may turn out to have an atrial myxoma. If you are a generalist physician, the odds are that, during your practice lifetime, you will, as I have, encounter a patient or two with Charcot-Marie-Tooth disease, central cord syndrome, Hansen disease, or myasthenia gravis (see Fig. 5.11). If a patient has a disease, that is the diagnosis, regardless of the odds.

Sotos, author of Zebra Cards, illustrates the risks of relying too heavily on the Woodward aphorism, describing the observations reported by a physician who took his daughter to the 1978 Worchester, Massachusetts Science Fair and Carnival: "We heard a sudden sound of hoof beats, followed by screams from the crowd. Expecting horses, I was surprised to see that two small zebras, pulling a cart, had gotten away from their handlers and were running amok" [3].

It seems in medical school we teach much more about the zebra diseases than about the more common "horse" ailments. But consider the lifeguard paradox: If lifeguards in training spent most of their time learning what they will later be doing most of the day, it would be honing their ability to apply suntan lotion to their own bodies.

1. Woodward T. Quoted in: Colgan R. Advice to the young physician. New York: Springer; 2009, pages 63–7.
2. Who coined the aphorism? Zebra cards. Available at: http://www.zebracards.com/a-intro_inventor.html.
3. Sotos JG. Zebra cards: an aid to obscure diagnosis. Mt. Vernon VA: Mount Vernon Book Systems; 1991.

The Patient's Story and the Good Medical History

A doctor who cannot take a good history and a patient who cannot give one are in danger of giving and receiving bad treatment.

Quoted by American cardiologist Paul Dudley White (1886–1973) [1]

Paul Dudley White, son of a family doctor, a graduate of Harvard Medical School, and a leading cardiologist of his day, cited the anonymous quote above in the Introduction to his 1955 book, *Clues In The Diagnosis And Treatment Of Heart Disease* (see Fig. 5.12).

There is an often-quoted adage "Listen to your patient; he (or she) is telling you the diagnosis." A well-constructed medical history is the result of a joint effort of the patient and physician. The facts provided by a patient are often presented in random order, and yet somewhere within them is there may be a cluster of vital clues. The physician's job is twofold. First, he or she must allow the patient to tell the story unprompted and without "jumping-in" with focused questions too early in the narrative, an exercise in restraint that can be difficult when the physician senses that the patient is wandering off course. The physician's second job is to herd the facts into a narrative that can lead to a useful differential diagnosis.

Fig. 5.12 Paul Dudley White. http://www.springer images.com/Images/ MedicineAndPublicHealth/ 1-10.1007_s00247-007- 0544-8-0

Here is an example: I once returned from vacation to learn that one of my patients had made several visits to see a resident in our clinic, complaining of dyspepsia and abdominal cramping. On the second visit, the resident had ordered diagnostic imaging studies, which were unrevealing. When I saw the patient a week later, I learned, as Paul Harvey would say, "the rest of the story" [2]. Her husband had lost his job, and they were in danger of losing their home. To make things worse, their teenage son had been suspended from school for fighting. Without this social history context, the history was incomplete. The patient had not fully told her story to the resident physician, whom she did not know well. The young doctor and the patient, together, had failed to construct the "good history" that would have explained the abdominal symptoms.

Howard Brody presents an interesting image of what happens in elicitation of the medical history: He holds that, metaphorically, the patient comes to the physician because the patient's narrative—his or her life story—is "broken." And the goal of the history phase of the medical encounter is to jointly construct a narrative that can respond to the plea that translates to: "My story is broken; can you help me fix it?" [3].

Sometimes the key fact that makes a seemingly-perplexing history come alive lies in the patient's answer to the question: "Can you tell me why you are here today; that is, what do you want from this visit?" I once had a patient—a middle-aged woman whom I had seen many times—come with a complaint of headache. What was curious was that she had had the same headache for decades and had never pursued this with me as a chief complain. Today, for some reason, the long-time recurring headache was a concern. When queried about the "real reason for the visit," she described how she had just received a note from her college roommate from years ago. The note told that the roommate had just been diagnosed with brain cancer, and was saying goodbye. At that moment the mist was lifted from the medical history, and what she wanted from the visit—reassurance rather than therapy—became clear.

1. White PD. Clues in the diagnosis and treatment of heart disease. Springfield IL: Charles C. Thomas; 1955, Introduction.
2. Harvey P. The rest of the story. New York: Bantam; 1984.
3. Brody H. "My story is broken; can you help me fix it?" Medical ethics and the joint construction of narrative. Literature and Med. 1994;13:79.

Chapter 6
Therapy and Healing

Pre - Islamic medicine

Therapeutics has come a long way since the gourd rattling and incantations of the earliest witchdoctors in prehistoric cultures. Their faith and that of their patients was in supernatural powers—sometimes gods of the sun or the moon, sometimes deities more ethereal.

Then came the discovery that some compounds had actual healing powers. Aloe, senna, castor oil, and opium were used in ancient Egypt centuries before the golden age of Greece. The bitter powder extracted from the bark of the willow tree and recognized by Hippocrates to combat pain we now know contained salicylates. Quinine, derived from the bark of the cinchona tree, was used to treat malaria in Rome as early as 1631 [1]. The foxglove plant was found in the late eighteenth century to have a component that could help combat dropsy, aka congestive heart failure, a factor we now call digitalis, the name derived from the "digits" of the foxglove plant [2] (see Fig. 6.1).

Of course there is more to therapy than medication. Brain surgery—using sharpened rocks—was performed as early as the Neolithic (late Stone Age) period. Stonecutters, those who operated to remove bladder stones, are mentioned in the classic Hippocratic oath. Cataract surgery was practiced as early as the fifth century BCE [3]. By the tenth century, surgery of the abdomen was being performed by the Arab physician Albucasis (936–1013) as described in his book *Al Tasrif* [4] (see Fig. 6.2).

Many of the favorable outcomes attributed to the rituals of the prehistoric times were undoubtedly the result of changes in the patients' outlook and psychological wellbeing. True psychiatric therapy came into its own with the work of Austrian neurologist Sigmund Freud (1856–1939), who introduced the concept of psychoanalysis and introduced us to the id, ego, and superego.

All this is intended as a prelude to the following pages that trace our thinking about therapy and healing from the time of Confucius in ancient China to the molecular therapy, microsurgery, and advances in treatment of mental illness that we currently employ.

1. Toovey S. The miraculous fever tree. New York: Harper Collins; 2004.
2. Wilkins MR et al. William Withering and digitalis, 1785 to 1985. Br Med J (Clin Res Ed). 1985;290:7.

© Springer Science+Business Media New York 2015
R.B. Taylor, *On the Shoulders of Medicine's Giants*,
DOI 10.1007/978-1-4939-1335-0_6

Fig. 6.1 Foxglove plant (Public domain). http://commons.wikimedia.org/wiki/Digitalis_purpurea#mediaviewer/File:Fingerhut_P5303439.jpg

Fig. 6.2 Portrait of Abu Alkasem AL Zehrawi (Albucasis). http://www.springerimages.com/Images/MedicineAndPublicHealth/1-10.1007_s00381-009-0912-9-0

3. Bellan L. The evolution of cataract surgery: the most common eye procedure in older adults. Ger Aging. 2008;11:328.
4. Fallouji MA. History of surgery of the abdominal cavity. Arabic contributions. Int Surg. 1993;78:236.

In Praise of the Old Remedies

Because the newer methods of treatment are good, it does not follow that the old ones were bad: for if our honorable and worshipful ancestors had not recovered from their ailments, you and I would not be here today.

Chinese philosopher Confucius (551–479 BCE) [1]

As I read the quotation above attributed to Confucius (see Fig. 6.3), I cannot help to recall the adage: Just because your patient gets better doesn't mean your medicine cured him. In fact, some of the pharmaceuticals administered to our honorable and worshipful ancestors were downright hazardous. Think of mercury as therapy

Fig. 6.3 Confucius. http:// www.springerimages.com/ Images/Environment/ 1-10.1007_978-1-4614- 3424-5_9-0

of syphilis, arsenic used to treat "herpetic affections" and as an ingredient in many patent medicines, and heroin marketed as an antitussive.

During my early practice days I saw a few patients with "phenacetin kidneys," renal toxicity caused by excessive use of phenacetin, notably in the form of APC tablets containing aspirin, phenacetin, and caffeine. Finally, in 1983, the US Food and Drug Administration ordered the withdrawal of phenacetin-containing drugs [2]. But, of course, the cohort of patients with phenacetin-induced nephropathy continued to be seen in physician's offices, including mine, for years.

On the other hand, are we sometimes too eager to try the new? This week I spoke with a friend in my community who described a cellulitis of the arm. He saw an infectious disease specialist and received an antibiotic prescription. The retail drugstore cost of the 2-week supply of the recently introduced antibiotic: $4,000. Might an older, and less costly, antibiotic have worked just as well?

In July 2012, the Food and Drug Administration approved the new drug Qsymia, a combination of phentermine and topiramate extended-release, for the "treatment of chronic weight management" [3]. Is this truly a great leap forward in the advancement of pharmacotherapy?

Robotic surgery of prostate cancer and other diseases has caught our fancy. And yet studies are showing that there may not be as many advantageous outcomes as hoped [4, 5].

In my practice I have always felt especially comfortable with remedies—whether pharmacologic or procedural—that are time-tested, that have a known and favorable risk/benefit ratio, and that are economical to use, whether the patient is paying directly or the cost is distributed to us all through some sort of insurance. In short, my favorite remedies are old, safe, and cheap.

1. Confucius. Quoted in: Strauss MB. Familiar medical quotations. Boston: Little Brown; 1968, p. 625.
2. Federal Register, October 5, 1983 (48 FR 45466).
3. FDA Approved Drugs. CenterWatch. Available at: http://www.centerwatch.com/drug-information/fda-approvals/default.aspx?DrugYear=2012.
4. Wood DP et al. Short-term health outcome differences between robotic and conventional radical prostatectomy. Urology. 2007;70:945.
5. Anderson JE et al. The first national examination of outcomes and trends in robotic surgery in the United States. J Am Coll Surg. 2012;215:107.

Being a Surgeon

Now a surgeon should be youthful with a strong and steady hand which never trembles, with vision sharp and clear, and spirit undaunted: filled with pity, so that he wishes to cure his patient, yet is not moved by his cries, to go too fast, or cut less than is necessary; but he does everything just as if the cries of pain cause him no emotion.

Roman physician and encyclopedist Aulus Cornelius
Celsus (ca. 25 BCE–ca. 50) [1]

De Medicina, the sole survivor of the written works of Celsus, contains eight books describing medical care in Rome during the first century. The work was apparently not well known during the time of Celsus—books copied by hand necessarily had a small audience. Lost for years, the book was found in a church in Milan in 1443, and subsequently was among the first published after the invention of the printing press in 1450. *De Medicina* contained many principles of medicine that endured for a millennium, and some descriptions that are taught today, such as

Fig. 6.4 Aulus Cornelius Celsus. http://www. springerimages.com/ Images/MedicineAnd PublicHealth/ 1-10.1007_978-1-4614- 0067-7_1-0

the cardinal signs of inflammation described by Celsus: *calor* (heat), *dolor* (pain), *rubor* (redness), and *tumor* (swelling) [1, 2]. Book VII of the treatise describes surgery (see Fig. 6.4).

Surgery is the most invasive of therapeutic interventions. I am a family physician, not a surgeon. I have, however, had a number of surgical experiences in my life. During training and my early practice years, I assisted on surgery of many types—abdominal, spinal, brain, vascular, and more. I have personally removed countless skin lesions and sutured hundreds of lacerations. Furthermore, I have personally undergone three major surgeries—involving lung, bowel, and hip. I think I know something about surgery.

As Celsus wrote, my ideal surgeon has a steady hand and clear vision. My surgeon should also have a bold spirit; let the non-surgeon ponder the self-confidence, perhaps bravado, needed to cut into another human being. Yet I also want a surgeon who empathically wants to heal me. Since the introduction of general anesthesia in the mid-nineteenth century, cries of pain are rarely heard in the operating suite. Yet I want the surgeon who has some emotion for the patient, definitely enough to avoid rushing, to cut just the right amount of flesh, and to manage postoperative recovery with skill and encouragement.

This brings me to the work of one of our contemporary surgeon-writers, Richard Seltzer, and his book *Mortal Lessons: Notes on the Art of Surgery*, a series of 19 essays on the practice of empathic surgery. Here, as an example, he describes the amputation of a gangrenous leg, one that he patiently has debrided for months, as he and the patient try to delay the inevitable. "At last we gave up, she and I. We could no longer run ahead of the gangrene. We had not the legs for it. There must be an amputation in order that she might live—and I as well. It was to heal both that I must take up knife and saw, and cut the leg off. And when I could feel it drop from her body to the table, see the blessed *space* appear between her and her leg, I too would be well." Seltzer goes on to tell that in the operating room he exposed the leg to find a message from the patient: a *happy face* and the words, "SMILE, DOCTOR" [3].

The three surgeons who operated on me—each quite different from the others— have all had the attributes described by Celsus, and the empathy we feel when reading Seltzer's stories. I did, however, not paint a clever message for any of them to discover in the operating room.

1. Celsus AC. *De medicina*, Book VII Praefat., first published in Florence, 1478. Trans Spencer WG.
2. Taylor RB. White coat tales: medicine's heroes, heritage and misadventures. New York: Springer; 2010, p. 10.
3. Seltzer R. Mortal lessons: notes on the art of surgery. New York: Harcourt Brace; 1974.

The Role of Natural Vitality in Healing

In treating a patient, let your first thought be to strengthen his natural vitality.

Persian physician and philosopher Muhammad ibn Zakarlya Razi,
known today as Rhazes (850–923) [1]

When the Dark Ages, with the medieval hyper-religiosity of the time, descended upon Europe, the epicenter of rationality and scholarship moved to the Middle East. One of the giants of this era was Rhazes of Persia, who wrote several hundred scientific papers and books, and is considered to have first described smallpox and differentiated it from measles. He also was an early champion of what today we think of as holistic medicine (see Fig. 6.5).

Fig. 6.5 Rhazes at the bedside of a child with measles. http://www. springerimages.com/ Images/MedicineAnd PublicHealth/1-10.1007_978- 1-4419-1034-9_3-0

Recognizing that the quotation is a translation, I nevertheless will focus on the word *vitality*. When I want to truly understand a word, I look to its origin: The word vitality comes from the Latin *vita*, meaning life. We have vital signs—blood pressure, pulse, respiratory rate, and temperature—all measuring functions essential to life. Blood is a vital fluid. I think that Rhazes connotes more than the simply somatic vigor. I sense an allusion to the patient's overall life force—one's physical, emotional, and spiritual wellbeing.

How can we, again as healers, help patients attain prompt and optimum wellbeing when there is an illness? One approach is creating a healing environment that focuses on the patient's overall life force. A few years ago I visited the local hospital in the small town of The Dalles (population 13,600), in Wasco County, Oregon. Here the patients not only had a hand in planning their diets, they chose the pictures to be hung on the walls of their hospital room and the music they would hear during the day—if they wished to listen to music. Patients had full access to their clinical records and to medical resources to learn about their diseases. Families were welcome at all hours. In short, the patients in this hospital received truly personal care, intended to help them achieve maximum wellbeing.

The hospital I had visited was an example of the Planetree model. Founded in 1978 and taking its name from the Plane tree (a type of sycamore, illustrated in Fig. 7.1) in whose shade Hippocrates taught on the Greek island of Kos, Planetree "is a nonprofit organization that provides education and information in a collaborative community of healthcare organizations, facilitating efforts to create patient-centered care in healing environments." It aims to empower families through information and education. A number of health care organizations in the USA and abroad have been designated as Planetree sites [2].

I happened to learn of this innovative approach to healing largely by chance. And why should I be impressed by encountering truly patient-centered care? Perhaps someday every patient will receive care intended to promote a full and prompt return to vitality in all spheres of life.

1. Rhazes. Quoted in: Walsh JJ. Medieval medicine. London: A & C Black; 1920.
2. Planetree. About us. Available at: http://planetree.org/?page_id=510.

Dogma and the Authority of the Physician

The art of medicine is not so fixed that we are without authority, no matter what we do; it changes according to the climates and according to the moons, according to Fernel and according to L'Escale (prominent physicians of the day). If your doctor does not think it good for you sleep, to drink wine, or to eat such-and-such a food, don't worry. I'll find you another who will not agree with him [1].

French essayist and statesman Michel de Montaigne (1533–1592)

De Montaigne, who wrote during the French Renaissance, is said to have influenced generations of authors who followed: Pascal, Emerson, Nietzsche, and Asimov. One of his memorable comments was *Que sais-je?*; this translates to "What do I know?" He might be considered a Renaissance "Doubting Thomas."

Fig. 6.6 Michel de Montaigne. http://www. springerimages.com/Images/ HumanitiesArts/1-10.1007_ s10539-011-9269-z-3

After all, the Renaissance was a time of questioning, of breaking away from medieval dogma, which was often rooted in religiosity. It was, perhaps, the earliest beginning of examining what we do (see Fig. 6.6).

Thus de Montaigne questioned "fixed" medical therapy of the day. Are abstinence from wine and the imposition of various dietary restrictions universally helpful in combating disease? Certainly, in the sixteenth century healers had no data-based evidence to guide them, but that did not prevent physicians, relying on tradition and experience, from offering their recommendations with considerable self-confidence.

What about today? In the twenty-first century, the era of science-based medicine, do we risk slipping back to dogmatic therapeutic recommendations, albeit based on published data? I am referring to clinical guidelines, promulgated by a wide variety of entities—quasi-governmental groups such as the US Preventive Services Task Force (USPSTF), specialty organizations such as the American College of Radiology (ACR) or the American Urologic Association (AUA), disease-specific organizations such as the American Diabetes Association (ADA), and many more. The National Guideline Clearinghouse (NGC) lists thousands of clinical guidelines originating in hundreds of various groups from the ACR (214 clinical guidelines) to the World Federation of Hemophilia (one clinical guideline) [2].

Should we be skeptical of clinical guidelines? Guidelines often disagree. The NGC lists prostate specific antigen (PSA) screening guidelines from four entities, including the USPSTF and the AUA. Qaseem et al. found these to be in conflict, which can confuse both patients and physicians [3]. Also, guidelines can be slow to change and are always based on retrospective data, if data-based at all [4]. The worst of all are GOBSAT guidelines—Good Old Boys Sat Around a Table.

In treating our patients, we should, of course, be aware of current clinical guidelines—even if they occasionally induce cognitive dissonance—while we seek the best therapeutic recommendations for our patients, based on evidence, experience, and the clinical context. In the words of de Montaigne, we are not "without authority."

1. de Montaigne M. Of experience (1587–88). Essays, Book III, Chapter 13. Trans. DM Frame. Roslyn NY: Walter J. Black, publisher; 1970.
2. Agency for Healthcare Research and Quality: National Guideline Clearinghouse. Available at: http://www.guideline.gov/browse/by-topic.aspx.
3. Qaseem A et al. Screening for prostate cancer: a guidance statement from the Clinical Guidelines Committee of the American College of Physicians. Ann Intern Med. 2013;158:1.
4. Ransohoff DF. How to decide whether a clinical practice guideline is trustworthy. JAMA. 2013;309:139.

About Making the Fewest Mistakes

> I have made mistakes myself; in learning the anatomy of the eye I dare say, I have spoiled a hatful. The best surgeon, like the best general, is he who makes the fewest mistakes.
>
> English surgeon and anatomist Sir Astley Paston Cooper (1768–1841) [1]

I am not sure that the "hatful" of eyes that Sir Astley Paston Cooper spoiled were anatomical specimens or those of patients. I hope it was the former. He was a leading surgeon of his day, famous for ligating an abdominal aortic aneurysm in 1817 and for removing an infected sebaceous cyst from the head of King George IV in

Fig. 6.7 Sir Astley Paston Cooper. http://www. springerimages.com/ Images/MedicineAnd PublicHealth/ 1-10.1007_978-1-84882-877-3_1-11

1820, the latter event probably a key factor in his subsequently being awarded the title of baronet. He did pioneering work on hernia surgery, and we remember him today as the namesake of the inguinal ligament of Cooper, aka the pectineal ligament (see Fig. 6.7).

Today, medical mistakes are a topic of keen interest. O'Connor et al. tell that "Studies estimate that a degree of error occurs in 5–15 % of all hospital admissions, with 45 % of errors occurring in the operating theater" [2]. Inpatient hospital medication errors are especially common. In one study in the intensive care setting using direct observation, there was one medication error for every five doses of medication administered [3]. The annual cost of measurable medical errors in the USA has been estimated to exceed $17 billion [4].

The medical error problem was thrust to the foreground in 2001 publication of *Crossing the Quality Chasm: A New Health Care System for the 21st Century*, a consensus report by the Institute of Medicine. Although this report covered a wide range of topics, the first of its six stated aims was "avoiding injuries to patients from the care that is intended to help them" [5].

What should we think of the adage by American surgeon J. Chalmers Da Costa (1863–1933): "What we call experience is often a dreadful list of ghastly mistakes?" [6]. Mistakes we make in practice, such as failing to notice a stiff neck in a febrile child, are learning experiences for the doctor, but are potentially consequential errors nevertheless. With the current increased attention to error prevention and to the quality of health care in general, we can hope that tomorrow's physician will gain experience without a series of "ghastly mistakes."

1. Cooper AP. The principles and practice of surgery (1824). Carlyle J et al. Fraser's Magazine. 1862;66:574.
2. O'Connor T et al. Safe surgery, the human factors approach. The Surgeon. 2010;8:93.
3. Kopp BJ et al. Medication errors and adverse drug events in an intensive care unit: direct observation approach for detection. Crit Care Med. 2006;34:415.
4. Van Den Bos J et al. The $17.1 billion problem: the annual cost of measurable medical error. Health Affairs. 2011;32:596.
5. Institute of Medicine. Crossing the quality chasm: a new health care system for the 21st century. Atlanta GA: National Academies Press; 2001.
6. Da Costa JC. The trials and triumphs of the surgeon. Pittsburgh: Dorrance Publishers; 1944, Chap. 1.

Confidence at the Operating Table

In a wound that is perfectly dry, and in tissues never permitted to become even stained by blood, the operator, unperturbed, may work for hours without fatigue. The confidence gradually acquired from the masterfulness in controlling hemorrhage gives to the surgeon the calm that is so essential for clear thinking and orderly procedure at the operating table.

American Surgeon William Stewart Halsted (1852–1922) [1]

Here I continue with the surgeon and surgery. Halsted must have had a large measure of the surgeon's confidence: at age 30, he performed a cholecystostomy

Fig. 6.8 William Stewart Halsted. http://www. springerimages.com/ Images/MedicineAnd PublicHealth/ 1-10.1007_s00268-010- 0580-7-4

(not a common operation at the time) on his own mother, on a kitchen table, in the middle of the night. He performed the USA's first radical mastectomy for breast cancer, and the procedure still carries his name. He was a champion of hemostasis in surgery—controlling bleeding during the procedure—and once wrote, "The only weapon with which the unconscious patient can immediately retaliate upon the incompetent surgeon is hemorrhage" [2] (see Fig. 6.8).

Here I want to focus not on bleeding and hemostasis but on the intensity involved in being a surgeon. There are, broadly speaking, two types of doctors: "surgical" and "medical." Future surgical doctors realize who they are, and are recognizable early in their medical school years, not by any superior skill with their hands—that will come later—but by a special decisive and straightforward approach to life [3].

In my online meanderings I came to an article published in *Dartmouth Medicine*, showing a series of paintings of life in the operating room by surgeon Joseph R. Wilder, with comments by former US Surgeon General C. Everett Koop. I will quote some passages here. About contemplation before surgery, Koop writes: "I know the surgeon has just scrubbed his hands and arms to the elbow and while doing so was subconsciously or consciously going over the details of the operation. His body and mind are now as prepared as they will ever be to engage the surgical challenge ahead" [4].

A few pages later, Koop tells, "Intense concentration is the hallmark of success-ful surgery" And Wilder himself believes "that the operating room should be seen as a cathedral, in which all the occupants must respect the sanctity of the place. All attention must be focused on the patient and the operative procedure .... A quiet ambience must be pervasive, with the same respect we show in our place of worship" [4]. This article describes a surgeon in whom I could have confidence.

1. Halsted WS. In: Bull John Hopkins Hospital. 1912;19:71.
2. Halsted WS. In: Bull John Hopkins Hospital. 1912; 23:191.
3. Taylor A. How to choose a medical specialty. Minneapolis: Mill City Press; 2012, page 328.
4. Wilder JR. The art of surgery. Dartmouth Medicine. Fall, 2002, page 28. Available at: http://dartmed.dartmouth.edu/fall02/pdf/Art_of_Surgery.pdf.

The Physician and Prescription Regret

A man who cannot work without his hypodermic needle is a poor doctor. The amount of narcotic you use is inversely proportional to your skill.

German-born American physician and author
Martin H. Fischer (1879–1962) [1]

Despite the obvious benefits in regard to pain relief, narcotic use never cured any disease. But opiate derivatives have caused problems for many, including some of our "medical giants." Sigmund Freud wrote a famous paper *Über Coca* in 1884, and engaged in "compulsive cocaine abuse" for 12 years (see Fig. 6.9). William Halsted struggled with addiction to cocaine and morphine throughout his adult life [2].

In July, 2013 the director of the US Centers for Disease Control and Prevention, Dr. Tom Frieden, declared that physicians are writing far too many narcotic prescriptions, "relying on these powerful drugs to treat chronic pain when physical therapy, exercise and other remedies would be safer and in many cases

Fig. 6.9 Cocaine powder (Public domain). http://commons.wikimedia.org/wiki/Cocaine#mediaviewer/File:Cocaine3.jpg

more effective." Drug overdose is one of the few causes of death that is rising in the USA [3].

According to the National Institute on Drug Abuse, of the reported seven million Americans who abuse prescription drugs, 5.1 million use opioids [4]. I am dismayed at the number of patients I have encountered who take huge doses of opioid analgesics every day for arthritis, back pain, and so forth, with no end in sight. Abuse extends to non-opioid depressants, notably alprazolam (Xanax) and diazepam (Valium); and stimulants such as amphetamine/dextroamphetamine (Adderall) and methylphenidate (Ritalin) [4].

The problem of excessive use of prescription drugs does not stop with opioids, depressants, and stimulants. Antibiotics are used for everything from the common cold to viral influenza—useless, of course, for both ailments.

Sixty years ago Finland and Weinstein raised the alarm over the widespread use of antibiotics and their "various untoward effects which have been increasing steadily; some are directly toxic, some are allergic, and others are related to the biologic activities of the chemotherapeutic substances" [5]. And in 1953, when the paper was published, there were only about a half dozen antimicrobials available.

Recently Sepkowitz has articulated our growing concern "about the risk of unleashing the doomsday organism" [6]. We are seeing multi-drug-resistant infections, such as tuberculosis and gonorrhea. Every time we prescribe an antibiotic unnecessarily, we risk the emergence of a few more resistant organisms.

In a busy clinic setting, the easiest thing to do is often to write a prescription. The patient is generally satisfied, and the visit is ended. But, like the narcotic overuse of old, it is often not the right thing to do.

1. Freud S. Über Coca. Therapie. 1884;2:289.
2. Markel H. An anatomy of addiction: Sigmund Freud, William Halsted, and the miracle drug cocaine. New York: Vintage; 2012.
3. Girion L et al. Doctors prescribe narcotics too often for pain, CDC chief says. Los Angeles Times. July 13, 2013. Available at: http://www.latimes.com/news/local/la-me-rx-painkillers-20130703,0,916397.story.
4. Sciuto L. Which prescription drugs do Americans abuse most? Available at: http://www.pbs.org/newshour/rundown/2013/04/which-prescription-drugs-do-americans-abuse-most.html.
5. Finland M et al. Complications induced by antimicrobial agents. N Engl J Med. 1953;248:220.
6. Sepkowitz KA. Finland, Weinstein, and the birth of antibiotic regret. N Engl J Med. 2012;367:102.

Becoming a Healer

It began to dawn on me that the healing art was not at all what people imagined it to be, that it was something simple, too simple, in fact for the ordinary mind to grasp. To put it in the simple way that it came to my mind I would say that it was like this. Everybody becomes a healer the moment he forgets about himself.

American author Henry Miller (1891–1980) [1]

Henry Miller was a "literary innovator" in many ways, including his development of the semi-autobiographical novel and his willingness to explore sexual topics [2]. *The Rosy Crucifixion*, a fictional autobiography cited above, was initially banned in the USA for its explicit sexuality, just one of his works to suffer this sanction.

His words inspired me to think about the word "healer." Of course, by no means are physicians the only healers. Probably every civilization has had its healers.

Fig. 6.10 Albrecht Dürer: "Christ among the Doctors", 1506. http://www.springerimages.com/Images/MedicineAndPublicHealth/1-10.1007_s00423-010-0673-7-12

Rarely did ancient medicine men and women "cure" any disease; but they often healed the patient's spirit, enough that peers valued their services. The primitive remedies used by the earliest physicians were, as in the case of bleeding and purging, sometimes more likely to cause harm, yet these ministrations had a "healing" action.

Jesus Christ was a healer. He and his followers happened to live in a time of terrifying pandemics, and the teachings of Jesus emphasized healing. In Luke 9: 1-2, the Bible tells that Jesus called his apostles and "He sent them to preach the kingdom of God and heal the sick." Cartwright theorizes that the remarkably rapid spread of Christianity in the time of Jesus and in the years following was due to a hope to escape disease. "Thus the growth of the Christian church was stimulated by its specific medical mission in a succession of plagues .... Conversions were numerous at all times of famine, earthquake or pestilences. At the height of the plague of Cyprian, the bishop and his fellow priests in North Africa were baptizing as many as two or three hundred persons a day" [3] (see Fig. 6.10).

Are we today seeing an increasing divergence between *healing* and *curing*? We read of cures little short of miraculous: Children with previously fatal childhood cancers may now be declared free of disease. Gene therapy has been used to cure chronic lymphocytic leukemia. Bariatric surgery has been described as a "surgical cure" for diabetes. But what about healing? As technology becomes increasingly dazzling and complex, the distance between the patient and physician often increases. Therapy is often prescribed, but then administered by a series of faceless technicians. Cures may occur, often leaving the patient feeling drained and depressed by the treatment, rather that exhilarated.

And so, what about *healing*? Increasingly, patients seek healing from nonphysicians. In the hospital cancer floors, counseling social workers are always busy. The *doula* has become an important part of childbirth for many women.

There are support groups for persons with almost any disease: heart disease, Huntington disease, Parkinson disease, celiac disease, and even peripheral neuropathy. Of course, in many cases, the healer is a family member. There is increasing recognition that anyone can become a healer "the moment he forgets about himself."

1. Miller H. The rosy crucifixion: Book one-Sexus (1949). New York: Grove Press; 1994; Vol 4, Chap. 14.
2. Sipper RB. Miller's tale: Henry hits 100. Los Angeles Times. Jan. 6, 1991.
3. Cartwright FF. Disease and history: The influence of disease in shaping the great events of history. New York: Crowell; 1972, page 23.

Understanding the Fundamentals of Therapy

One cannot possibly practice good medicine and not understand the fundamentals underlying therapy. Few if any rules for therapy are more than 90 per cent correct. If one does not understand the fundamentals, one does more harm in the 10 per cent of instances to which the rules do not apply than one does good in the 90 per cent to which they do apply.

American endocrinologist Fuller Albright (1900–1969) [1]

A Harvard Medical School graduate who spent much of his career at Massachusetts General Hospital, Fuller Albright was a highly productive investigator whose achievements included the description of polyostotic fibrous dysplasia, also known as McCune-Albright syndrome. His life took a sad turn when he developed Parkinson disease at age 37, a condition that prompted him in 1956 to undergo experimental brain surgery. Postoperative hemorrhage left him in a vegetative state from that time until his death in 1969 [2]. Is it not ironic, given the quotation above about "fundamentals of therapy," that Fuller's life was cut short by a therapeutic complication? (see Fig. 6.11).

Fuller's words brought to mind the topic of continuing medical education (CME), especially as it applies to therapy. Today, no physician believes that medical

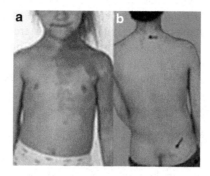

Fig. 6.11 McCune-Albright syndrome, with typical skin lesions, A and B. Source: Dumitrescu CE et al. McCune-Albright syndrome. Orphanet J Rare Dis. 2008;3:12. This file is licensed under the Creative Commons Attribution 2.0 Generic license. http://commons.wikimedia.org/wiki/File:Mccune-albrightsyndrome1.jpg

education ends upon receiving a diploma on graduation day. Not only are many new drugs, techniques, and therapeutic modalities introduced each year, whole new categories of entities emerge. When I attended medical school a half century ago, there were no beta-blockers, no calcium channel blockers, and no angiotensin-converting enzyme inhibitors. In fact scientists had not yet discovered beta-receptors, calcium channels, or angiotensin. I learned about all of these entities—and much more—after completing medical school and training.

I believe that the CME imperative is especially important as regards therapy. Why? Diseases don't change very much, and hence new information regarding diagnosis is small, indeed, when compared with advances in therapy, with new information coming daily.

With each new drug advance, it is important to discern *how the drug works*. For some of us, this calls for a refresher session on basic physiology. For example, when sumatriptan (Imitrex) was introduced to treat migraine, it was the first of its class and the first breakthrough in migraine headache therapy in a long time. But how did it work? Sumatriptan, as it turned out, is chemically similar to serotonin (5-hydroxytryptamine) and fundamentally has a similar action—vasoconstriction. By knowing this, I begin to understand its action, and I also can see that I should not prescribe the drug for persons with basilar or hemiplegic migraine, which would carry a risk of stroke.

There is great danger in learning therapeutic algorithms and following them in the clinic without knowing the underlying physiology and pharmacology. Only by understanding how drugs work and how the body handles them can you anticipate untoward effects and interactions. Recently there have been dozens of drugs released. Canagliflozin (Invokana) for the treatment of type II diabetes mellitus, ospemifene (Osphena) to treat vulvovaginal atrophy and dyspareunia, and budesonide (Uceris) for the management of ulcerative colitis are just a few. As physicians, neither you nor I should write our first prescription for one of these new medications until we have identified its class of drugs and could explain to a colleague the fundamental actions of drugs in this class.

1. Albright F (c. 1960). Quoted in: Introduction to diseases of the ductless glands. Textbook of medicine, ed. 8. Cecil RL, Loeb RF, editors. Philadelphia: Saunders; 1995.
2. Kleeman CR et al. Fuller Albright: the consummate clinical investigator. Clin J Am Soc Nephrol. 2009;4:1541.

Choices at the End of Life

> If (a doctor) goes on prolonging a life that can never again have purpose or meaning, his
> kindness becomes a cruelty .... We shall have to learn to refrain from doing things merely
> because we know how to do them. In particular we must have the courage to refrain from
> buying patients' lives at a price they and their friends do not want to pay.
>
> British physician and medical editor Sir Theodore Fox (1899–1989) [1]

Sir Theodore Fox, editor of *The Lancet* from 1944 to 1964, knighted in 1962, was invited to present the Harveian oration to the Royal College of Physicians of London, Oct. 18, 1965. In this oration he observed: "Life is not really the most important thing in life" [1].

Fox reminds that our job is to prolong life, but not to prolong dying. In the course of many illnesses there is a tipping point when the change occurs. Sometimes it is the event of intubation, or of using pressor drugs to artificially maintain failing organ perfusion. Sometimes the point when we moved from rational therapy to prolongation of dying can be discerned only in retrospect. In speaking with many families of critically ill persons, I have found the "prolonging life vs. prolonging death" dichotomy very useful in decision-making.

This brings me to the topic of "right-to-die" legislation. I practiced medicine in Oregon for 28 years, during the gestational years of the Oregon Death with Dignity Act, a controversial 1994 law allowing physicians, following a mandated series of steps, to prescribe lethal doses of drugs to terminally ill patients.

The popular support for legislation was bolstered by the words of the widow of Oregonian Emerson D. Hoogstraat: "In his final months, Emerson lived in agony, unable … to end his own suffering. His bones became so brittle that they broke when he turned over. He lived in constant pain, no matter how much morphine was prescribed. My husband of 40 years died exactly the death he feared" [2].

Today, the option of physician-assisted suicide is legally available in four US states and several European countries [3].

At times we need to reflect on what it means to live longer. When is the quality of life really so poor that ending it is a reasonable option? Agonizing pain seems to many a reasonable justification. How about extreme disability—such as quadriplegia? What about advanced, debilitated old age? Who decides? And who influences decisions? What is the role of the physician, who balances the innate desire to do good for the patient (beneficence) and to allow patients to make their own choices regarding what happens to their bodies (autonomy) with the admonition to "do no harm" (nonmaleficence)? And what about the physician's own beliefs? These questions are not really new, and will continue to challenge ethicists, and all of us, for a long time to come.

1. Fox T. Purposes of medicine. The Harveian oration for 1965, delivered before the Royal College of Physicians of London. The Lancet. 1965;286:801.
2. Emerson Hoogstraat helped pass the Oregon Death With Dignity Act. Oregon: Oregon Secretary of State Elections Division; 1997. Available at: http://www.sos.state.or.us/elections/nov497/voters.guide/M51/M51ao.htm.
3. Hendry M. Why do we want the right to die? A systematic review of the international literature on the views of patients, carers, and the public on assisted dying. Pall Med. 2013;27:13.

Chapter 7
Learning and Teaching Medicine

What institution can claim to be the first medical school?

The medical school in Montpellier, France, is described as the world's oldest medical school in continuous operation. It dates to the year 1220, when it was chartered by the Catholic Church, and today has approximately 7,000 students, which seem to me to be a very large student body for a medical school. Physician and astrologer Nostradamus studied at this school in the early sixteenth century [1].

In 1222 the University of Padua in Italy began as a school of law. Its venerable school of medicine, with anatomist Andreas Vesalius (1514–1564) as its shining light, began some years later [2]. Aberdeen University Medical School in Scotland, dating to 1495, has been described as the first medical school in the English-speaking world [3].

Originally founded in 1765 as the Medical Department of the College of Philadelphia, the University of Pennsylvania School of Medicine in Philadelphia is America's oldest medical school. Doctor Benjamin Rush (1745–1813), who signed the US Declaration of Independence, taught clinical practice there [4]. The University of Maryland School of Medicine in Baltimore, founded in 1807 with John Beal Davidge (1768–1829) as first dean, holds the distinction of being the oldest public medical school in the USA [5].

We believe that Hippocrates taught medicine to his pupils under the plane tree in the center of town on the Greek island of Kos. Visitors today can see the current tree, by legend about five centuries old, but probably actually growths descended from the original "tree of Hippocrates" [6]. Seeds from this plane tree were planted on the grounds of the US National Library of Medicine in 1961 [6]. The "tree of Hippocrates" is, of course, the namesake of the Planetree movement, described in Chap. 6 (see Fig. 7.1). Of course, there was certainly more or less formal teaching in ancient China and India predating all the above.

© Springer Science+Business Media New York 2015
R.B. Taylor, *On the Shoulders of Medicine's Giants*,
DOI 10.1007/978-1-4939-1335-0_7

Fig. 7.1 The "tree of Hippocrates" on the island of Kos (Public domain). http://commons. wikimedia.org/wiki/Tree_of_Hippocrates#mediaviewer/File:Plane_tree_of_Hippocrates.jpg

By no means does all medical teaching and learning take place in classrooms and teaching hospitals. Some of our best educational experiences take place at the bedside in community hospitals and in the private offices of physicians who volunteer to teach tomorrow's doctors.

This chapter presents the words of some of the giants who have influenced medical education, and shows ways in which their wisdom has shaped how we teach and learn medicine today.

1. Which medical school is the world's oldest? Everwell. Available at: http://www.everwell.com/insights/health_hits/world_oldest_medical_school.php.
2. Universita degli studi di Padova. Available at: http://www.unipd.it/coimbrameeting2011/university/university_history.html.
3. Carter J. Crown and gown: illustrated history of the University of Aberdeen, 1495–1995. Aberdeen: Aberdeen University Press: 1994.
4. What are the oldest medical schools in the US? Available at: http://wiki.answers.com/Q/What_are_the_Oldest_medical_schools_in_the_U.S.
5. University of Maryland School of Medicine: our history. Available at: http://medschool.umaryland.edu/history.asp.
6. Kos Hippocrates plane tree. Available at: http://www.greeka.com/dodecanese/kos/kos-excursions/kos-hippocrates-plane-tree.htm.

Learning While Teaching

Associate with those who will make a better man of you. Welcome those whom you yourself can improve. The process is mutual; for men learn while they teach.

Roman philosopher and statesman Lucius Annaeus Seneca (c. 4 BCE–65 AD) [1]

The quotation above is taken from a series of letters written by Lucius Annaeus Seneca, aka Seneca the Younger, while in exile in Corsica after being banished from Rome by the Emperor Claudius based on a charge of adultery. The letters were sent to Lucilius, the Governor of Sicily, and the topics were wide ranging: the sanctity of marriage, the "equality" of slave and master, the value of learning, the merits of engaging reflectively with our lives, and on the principles of Stoicism [2].

Seneca understood one of the key rewards of teaching—the teacher learns. The best teachers share certain attributes: They are *enthusiastic;* that is, they are passionate about their subject. A professor of dermatology must believe it is vitally important that each student be able to recognize psoriasis and know the spectrum of therapy for atopic dermatitis. They *understand educational principles,* such as insisting that in the clinical teaching setting, it is pedagogically sound for the student to make a commitment to a proposed diagnosis or therapy, a declaration which the instructor may elaborate upon or, if not on target, offer correct information. They *know what methods work best*—whether lecture, small group discussion, problem-based instruction, or one-on-one clinical teaching—according to the topic, the setting, and the learners. And they *elicit feedback,* so that they keep improving as teachers.

There is one more characteristic of the best teachers, one that is seldom articulated: They love to learn more about medicine. This is fortunate, because, as Doctor Charles H. Mayo (1865–1939), cofounder of the Mayo Clinic in Rochester,

Minnesota, once stated: "Once you start studying medicine, you never get through with it" [2]. And there may be no better way to learn than to teach.

Reading journals and books, attending lectures, and learning on the job from colleagues are all very good. But the best way to master a topic may just be to teach it. One possibility is to accept a student into your office for a clinical preceptorship or clerkship. The Family Medicine Department at the medical school in Oregon has over 200 board-certified family doctors on our volunteer faculty. These clinical faculty members enthusiastically welcome students into their offices, and sometimes their homes. At the end of the academic year, comments from these community-based educators often are expressed as some variation of "I think I learned more than the student." As a practicing physician, there is nothing like having a bright young learner at your side, someone who might have heard a lecture last week on the very disease affecting the patient in front of you.

Another way to learn while teaching is to present a seminar to students, residents, or colleagues on a slightly unfamiliar topic, yet one related to your specialty. A cardiologist may present "Dietary Approaches to the Prevention of Coronary Artery Disease," or an obstetrician/gynecologist might speak on "Psychologic Implications of Female Infertility." Just as likely to be a learning opportunity is to write an article or book chapter. In your effort to pass the journal's peer review process, and to avoid inflammatory letters to the editor, you are sure to research your topic thoroughly, choose each word with great care, and learn more than you knew before writing.

As French essayist Joseph Joubert (1754–1824) once remarked: "To teach is to learn twice" [3].

1. Seneca LA. Moral letters to Lucilius—letters from a stoic. Letter VII: On Crowds, line 8. In: *Les Belles Lettres*. Noblot H, translator. Paris: *Société d'Édition*; 1941: page 21.
2. Mayo CH. In: Aphorisms of Dr. Charles Horace Mayo and Dr. William James Mayo. Willius FA, editor. Rochester MN: Mayo Foundation for Medical Education and Research; 1988.
3. Joubert J. *Pensées* (Thoughts, published posthumously). Paris: Chateaubriand; 1838.

About the Lecture in Teaching and Learning

> Here is a great hospital; and here I hold that all teaching by lectures should have for its first
> and principle purpose to give effect to that self-teaching, which, from the objects which
> surround us, all may practice and profit by who have eyes and ears and a docile mind. Do
> not believe a word that I say until you have gone into the wards and proved it. There you
> will find your great book of instruction. I only pretend to supply a key, a glossary, or an
> index to it. Use that book as you ought, and then, though in the end you and I may have the
> same knowledge, it will not be because it has passed from my mind to yours, but, being
> gained by your own observation, ratified by your own proofs, and matured by your own
> thought, you will have and hold it as your own independent possession.
>
> British physician and medical educator Peter Mere Latham (1789–1875) [1]

Peter Mere Latham was a leader in medical education in the early nineteenth
century, and a prescient one, at that. Latham was one of the first to adopt Laennec's

Fig. 7.2 What is the future of the classroom lecturer with a PowerPoint presentation? http://www.
springerimages.com/Images/HumanitiesArts/1-10.1007_s11948-011-9295-x-3

stethoscope. He advocated "examination of the bodies of those who die" as a way to understand disease. He believed in relating murmurs heard to specimens of heart valves. Latham had "long doubted whether systematic courses of lectures on medicine or surgery ought to be considered as essential a part of medical education as they are" [2]. His concept of the primacy of clinical instruction over lectures seemed to anticipate Sir William Osler, who wrote a half century later, "The best life of the teacher is in supervising the personal daily contact of patient with the student in the wards," and, more to the point, "Superfluity of lecturing causes ischial bursitis" [3] (see Fig. 7.2).

The rise of problem-based learning (PBL), beginning in the late 1960s, reflected dissatisfaction with the "sage on a stage" method of teaching in most medical schools at the time. PBL was originally championed at McMaster University School of Medicine in Hamilton, Ontario, Canada, and spread across North America. Various iterations of PBL continue in many medical schools today.

Is the lecture an anachronism in medical school education? Not really. It is being saved by technology, but with a change in the job description for many medical school instructors. Prober and Heath propose "a flipped classroom model, in which students absorb an instructor's lecture in a digital format as homework, freeing up class time for a focus on applications, including emotion-provoking simulation exercises …. Teachers would be able to actually teach, rather than merely make speeches" [4].

Even in instances when lectures are considered the best way to communicate information, do we need each instructor to create his or her own lecture? According to University of Maryland physics professor Joe Redish, "With modern technology, if all there is is lectures, we don't need faculty to do it. Get 'em to do it once, put it on the web, and fire the faculty" [5]. In this model, the nation's few best medical school lecturers will be franchised on line; the rest will go back to their laboratories or teach while caring for patients in the clinic or hospital.

1. Latham PM. Diseases of the heart, Philadelphia: Barrington & Haswell; 1847.
2. Spaulding WB. Peter Mere Latham (1789–1875): a great medical educator. Can Med Assoc J. 1971;194:1109.
3. Osler W. Quoted in: Silverman ME et al. The quotable Osler. Philadelphia: Am Coll Phys; 2003, page 217.
4. Prober CG et al. Lecture halls without lectures—a proposal for medical education. N Engl J Med. 2012;366:1657.
5. Don't lecture me: rethinking how college students learn. MindShift. February 20, 2012. Available at: http://blogs.kqed.org/mindshift/2012/02/dont-lecture-me-rethinking-how-college-students-learn-2/.

Always Being a Student

You must always be students, learning and unlearning till your life's end, and if, gentlemen, you are not prepared to follow your profession in this spirit, I implore you to leave its ranks and betake yourself of some third-class trade.

British surgeon Lord Joseph Lister (1827–1912) [1]

Lister is a good exemplar of standing on the shoulder of giants. His well-recognized contributions to surgical antisepsis built upon the work of von Leeuwenhoek, who observed "animalcules" through his microscope (1764); Semmelweis, the misunderstood champion of hand-washing to prevent puerperal fever (1850); and Pasteur, who theorized that the causes of contagious diseases were like the causes of fermentation (1859). Then, in 1865, Lister began using an antiseptic (carbolic acid, aka phenol) to prevent infection in the surgical field [2] (see Fig. 7.3).

The concept of learning (and sometimes unlearning) until life's end—or at least until the end of an active career—is the foundation of Continuing Medical Education (CME), a major industry in America. One reason for the success of the CME enterprise,

Fig. 7.3 Lord Joseph Lister.
http://www.springerimages.
com/Images/Medicine
AndPublicHealth/1-10.1007_
s00129-003-1422-6-2

aside from the physician's appetite for current knowledge, is the requirement that CME must be documented before state medical licenses will be issued and renewed. CME activities include workshops, lectures, video and audio presentations, online learning including presentations by "rock star" lecturers (described above), reading scientific journals and books, and attending medical meetings.

Over the years, medical meetings have been a source of up-to-date knowledge, a chance to see old friends, and, at times, an excuse to take the family on vacation. Medical meetings, whether in a nearby lecture hall or Cancun, are sometimes where groundbreaking advances are first described as research reports. Getting away from home, often with family members, and seeing colleagues from other locations are among the side benefits of CME events.

But there is a dark side. In the 1970s and 1980s, physicians were enticed to attend elaborate industry-sponsored CME meetings, often with lectures by carefully chosen "experts," held all-expenses-paid at prime vacation sites. I now confess speaking at one such meeting at a luxury hotel in Palm Springs, California, with the final dinner held in a huge white tent erected in the desert, featuring a lavish candle-light dinner and an orchestra for dancing. That was long ago. Scrutiny by the Accreditation Council for Continuing Medical Education (ACCME) and other groups has stopped these junkets—in the USA [3]. But not elsewhere. The August 19, 2013 edition of the Wall Street Journal described all-expense trips for Chinese doctors to premier tourist destinations, including Turkey and Istanbul, billed as CME, and intended to persuade the doctors to prescribe the sponsoring company's medications [4].

We have made strides, but should not be complacent. The pharmaceutical industry still supported 31 % of the US continuing medical education activity in 2010 [5]. Lifelong learning, as espoused by Lister, is part of being a physician, but we must assure that what we learn in continuing education is not only current, but also unbiased and free of inducements to favor one company's product.

1. Lister J (c. 1870). Quoted in: Strauss MB. Familiar medical quotations. Boston: Little Brown; 1968, page 261.
2. Germ theory calendar. Available at: http://germtheorycalendar.com/db.aspx.
3. Steinman MA et al. Industry support of CME—are we at the tipping point? N Engl J Med. 2012;366:1069.
4. Glaxo junkets highlight ills of Chinese medicine. Wall Street Journal. August 19, 2013; page B1. Available at: http://www.fiercepharma.com/story/wsj-glaxosmithkline-payments-junkets-lapped-underpaid-chinese-doctors/2013-08-19.
5. Lexchin J et al. Should the C in CME stand for commercial? So Med J. 2012;5:3.

Fewer but Better, but for Whom?

> If the sick are to reap the full benefit of recent progress in medicine, a more uniformly arduous and expensive medical school education is demanded.
>
> American educator Abraham Flexner (1866–1959) [1]

At the beginning of the twentieth century in America, medical education often involved either apprenticeship training or enrollment in one of the proprietary medical schools whose chief criterion for admission was the ability to pay tuition. This all changed following the 1910 publication of the "Flexner Report" [1].

Flexner was an educator, and not a physician. In 1908 he published a report titled *The American College*, critical of higher education in the USA [2]. Based on this work, the Carnegie Foundation for the Advancement of Teaching engaged Flexner to produce an analysis of American medical education (see Fig. 7.4).

Fig. 7.4 Abraham Flexner. http://www.springerimages. com/Images/Medicine AndPublicHealth/1-10.1007_ s40037-012-0002-7-0

Flexner undertook his task with gusto. He made a whirlwind tour of US medical schools, visiting 155 institutions in 98 cities. In a single month in 1909, he visited 30 schools in 12 cities. His report, published in 1910, called for fewer but better medical schools, and recommended closing 75 % of existing US medical education institutions [3].

Many of Flexner's calls for improvements in medical education were timely and logical. But some have had profound unanticipated effects. Reflecting the German idea of the time that medicine should be studied as a laboratory science, Flexner proposed that those teaching medical students should be salaried physicians with a "research function." This was an elitist notion that sparked the town-gown discord often seen today. The proposed exclusion of practicing clinicians from medical education was strongly opposed by Sir William Osler, who wrote in 1911, "I cannot imagine anything more subversive to the highest ideal of the clinical school than to hand over our young men who are to be our best practitioners to a group of teachers who are ex-officio out of touch with the conditions under which these young men will live" [4].

There has been one other important outcome of the changes made, mentioned in the quote above. Medical education in the Flexnerian model has become very expensive. Prior to 1910, farm boys (It was pretty much all men at that time.) could learn their trade as apprentices and then become general physicians serving their communities. Now the average annual tuition for at both public and private medical schools is more than $40,000. Add in fees, health insurance, and living costs, and we see why young doctors graduate with an average debt exceeding $162,000 [5]. Have we created a medical education system open only to the affluent or to those who can eventually secure a residency position in a lucrative specialty?

1. Flexner A. Medical education in the United States and Canada. New York: Carnegie Foundation for the Advancement of Teaching; 1910.
2. Flexner A. The American college: a criticism. New York: The Century Company; 1908.
3. Markel H. Abraham Flexner and his remarkable report on medical education: a century later. JAMA. 2010;303:888.
4. Osler W. On full-time teaching in medical schools (1911). Reprinted in: Can Med Assoc J. 1962;87:76.
5. U.S. needs thousands of doctors while average medical student faces loan debts in excess of $162,000. Available at: http://www.kshb.com/dpp/news/national/US-needs-thousands-of-doctors-while-average-medical-student-faces-loan-debts-in-excess-of-162000.

About Teaching and Medical Fads

Those of us who have the duty of training the rising generation of doctors…must not inseminate the virgin minds of the young with the tares of our own fads. It is for this reason that it is easily possible for teaching to be too "up to date." It is always well, before handing the cup of knowledge to the young, to wait until the froth has settled.

Scottish physician and author Sir Robert Grieve Hutchison (1871–1960) [1]

Hutchison was the author of the book *Clinical Methods*, which was first published in 1897 and is currently in its 23rd edition [2]. We also recall Sir Robert as the eponymous namesake of Hutchison disease, a malignant tumor typically found in children, passed by autosomal recessive inheritance, sometimes associated with assorted congenital malformations, and often spreading to bones, liver, and lungs.

Hutchison had a gift for clear and vivid communication. In his discussion of medical fads, Hutchison writes: "… the ghosts of dead patients that haunt us do not ask why we did not employ the latest fad of clinical investigation. They ask us, why

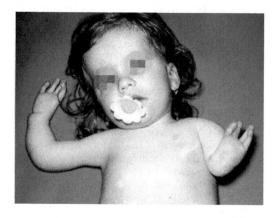

Fig. 7.5 Child with left-sided phocomelia. The forearm attaches directly to the shoulder. http://www.springerimages.com/Images/MedicineAndPublicHealth/1-10.1007_978-3-540-85561-3_15-0

did you not test my urine?" Hutchison favored using sound, time-tested clinical principles over being the early adapter of new trends in medicine.

What about the fate of some historic fads? In the 1940s premature infants often were treated with high-dose oxygen therapy, found subsequently to be a cause of retrolental fibroplasia/retinopathy of prematurity. In the 1950s medical students were taught to use diethylstilbestrol (DES) to reduce the risk of pregnancy complications—until we learned that girls and women exposed to DES were at later risk for rare vaginal tumors. In the early 1960s thalidomide was routinely prescribed for pregnant mothers complaining of morning sickness, resulting in many cases of birth defects such as phocomelia (see Fig. 7.5).

Other medical fads have included the treatment of sore throats with silver protein, still available without prescription, despite the occasional case of argyria. We once treated peptic ulcer disease with a Sippy diet, and hypertension with phenobarbital. Today we are prescribing large doses of vitamin D for our patients and treating many school-age children with stimulants to control behavior attributed to attention-deficit/hyperactivity disorder.

Will evidence-based medicine protect us from medical faddism, from being too eager to try the new? What about oxygen therapy of premature infants, DES, and thalidomide? It took years to learn the long-term hazards of these apparently "safe" treatments. Clinical studies at the time of their popularity could only document their effectiveness, not the long-term risks. In the case of varenicline (Chantix), a 2006 evidence-based study showed the drug's effectiveness [2]. It was after the drug was released to the public that a post-marketing study showed that it may cause erratic and suicidal behavior.

Hutchison holds that "as individuals we rise on the stepping-stones of our dead fads to higher things" [1]. We just need to be careful that we do not pass potentially harmful fads to our students as "facts."

1. Hutchison R. On fashions and fads in medicine. BMJ.1925;1/3361:995.
2. Jorenby DE et al. Efficacy of varenicline, an alpha4beta2 nicotinic acetylcholine receptor partial agonist, vs. placebo or sustained-release bupropion for smoking cessation: a randomized controlled trial. JAMA. 2006;296:56.

Being a Teacher and a Student

The safest thing for a patient is to be in the hands of a man engaged in teaching medicine. In order to be a teacher of medicine the doctor must always be a student.

America physician Charles Horace Mayo (1865–1939) [1]

The beginning of the Mayo Clinic in Rochester, Minnesota, is often listed as 1889, the date of the opening of Saint Mary's Hospital. This world-renowned institution grew from the frontier practice of Doctor William Worrall Mayo, later joined by his sons William James Mayo (1861–1939) and Charles Horace Mayo (see Fig. 7.6).

Fig. 7.6 Charles Horace and William James Mayo. http://www.springerimages.com/Images/ MedicineAndPublicHealth/1-10.1007_s00105-003-0675-2-4

From the beginning, the brothers stressed the importance of education as part of patient care. The brothers helped found the University of Minnesota Medical School in 1917 and, later, in 1972, the Mayo Medical School was opened [2].

The Brothers Mayo seemed, like Osler, not to embrace the Flexnerian model of limiting medical teaching to full-time academicians with a "research function" [3]. The early Mayo Clinic years involved teaching medical students and young physicians by *experienced practicing physicians* [2]. Today, the Mayo Clinic Hospital is world renowned for excellence in patient care, ranking the third best overall hospital nationally in 2013 by US News & World Report [4].

Which brings me to the message I glean from the quote above: The best doctors are those who have integrated teaching into their practices. They thus must know the latest clinical research findings, because their students will have found them online. And they must be prepared to answer questions such as the following: "Why do you do it this way?" and "Why did you choose this drug and not that?" A reply that "We have always done it this way" will not suffice.

Then there is the question: Who is a teacher of medicine? I have heard it said that "there is no good teaching outside the medical school." But at many medical schools, clerkships with community physicians, especially rural physicians, receive the highest ranking by students. The presence of learners challenges the community doctor to "always be a student."

Students demand value for their tuition dollars. If they were not receiving valuable education, these community preceptors would not continue to receive high marks. And, with students present, the patient care provided by these clinician-teachers is better than ever, even if these physicians are not full-time medical school employees.

1. Mayo CH. Quoted in: Proceedings of the Staff Meetings of the Mayo Clinic, 1927.
2. Mayo Clinic Tradition and Heritage. Available at: http://www.mayoclinic.org/tradition-heritage/brothers-practice.html.
3. Flexner A. Medical education in the United States and Canada. New York: Carnegie Foundation for the Advancement of Teaching; 1910.
4. Best hospitals 2013–2014: Overview and honor roll. US News & World Report. Available at: http://health.usnews.com/health-news/best-hospitals/articles/2013/07/16/best-hospitals-2013-14-overview-and-honor-roll.

Teaching the Teachers

The preceptor, trying to instill educative principles in his students, cannot foretell how far a beam his tiny light will throw. Aesculapius couldn't guess; Hippocrates couldn't guess; neither can he. But it is probable that some of the medical students he educated to teach will, in life, have their eyes more lifted to the stars and later find a grateful, smiling Hippocrates waiting to greet them in some medical Elysium.

American physician and educator David Seegal (1899–1972) [1]

There are multiplier effects with each generation of descendants. For example, it has been reported that there are hundreds of millions of descendants of William the Conqueror of England, more than the current population of the UK, Canada, the USA, and Australia combined [2]. This somewhat whimsical factoid illustrates the ripple effect that can occur over generations. This sort of effect also occurs with the influence of an educator, and especially one who persuades his or her pupils to *teach.*

For example, I have been a medical educator for more than three decades. The students and residents I have taught have, for the most part, gone on to productive medical careers, each caring for hundreds, perhaps thousands of patients each year. I like to think that today, somewhere in America, a practicing physician may recall a physical exam maneuver or a therapeutic pathway based on my teaching. I recall the words of Osler: "… no bubble is so iridescent or floats longer than that blown by the successful teacher" [3].

Even greater multiplier potential attends those who educate the students who themselves become teachers, whether in academic medical centers or in a community setting. Hippocrates was certainly one of these, as were Galen, Maimonides, Cushing, Osler, Woodward, and others. They taught and their pupils then went on to teach. Not all their student progeny necessarily taught in the classical sense of lecturing in the classroom. Many wrote scholarly articles and books; others influenced young minds as role models at the bedside or in the clinic.

Are we doing what we should to teach our physicians how to teach? A survey of 108 teaching hospitals regarding faculty development activities led to the following conclusion: "A minority of US teaching hospitals offer ongoing faculty development (FD) in teaching skills. Continued progress will likely require increased institutional commitment, improved evaluations, and adequate resources, particularly FD instructors and funding" [4].

We may actually be doing more to help medical students learn to be teachers. A 2008 survey of US medical schools revealed that 43 of 99 responding schools offer formal students-as-teachers programs, generally in the senior year of medical school [5].

Each of us practicing and, we hope, teaching medicine can trace some thread of our knowledge to the great physician-educators, who had "their eyes more lifted to the stars." Today the intellectual descendants of teachers like Hippocrates, Galen, Maimonides, and even some more recent giants—taken together—must surely outnumber those of the Norman conqueror of England.

1. Seegal D. Teaching medical students to teach. J Med Educ. 1964;39:1030.
2. Yahoo! Answers. Available at: http://answers.yahoo.com/question/index?qid=20101207220850AA2zZfd.
3. Osler W. The pathological institute of a general hospital. Glasgow Med J. 1911;76:321.
4. Clark JM et al. National survey of faculty development in departments of medicine of US teaching hospitals. J Gen Int Med. 2004;19:205.
5. Soriano RP et al. Teaching medical students how to teach: a national survey of students-as-teachers programs in US medical schools. Acad Med. 2010;85:1725.

Aspiring Tailors and Physicians

Medical education has been compared to teaching would-be tailors about the molecular structure of wool, how to grow cotton, and how yarn is spun in a mill. After that, they are sent out to tailor clothes.

American physician Howard M. Spiro (1924–2012) [1]

Howard M. Spiro, founding chief of the Yale School of Medicine section on gastroenterology, was a remarkable physician. When he relinquished this post in 1982 following a tenure of 27 years, he became director of the Yale Program for Humanities in Medicine. What shall we make of Spiro's metaphorical depiction of contemporary medical education?

Spiro seems to be describing the "basic sciences" of tailoring. Generations of medical students, eager to survive the basic sciences on their way to future clinical rotations, have memorized the Krebs cycle, identified in 1937 by Nobel laureate Hans Adolf Krebs. It describes how energy is derived from the oxidation of sugars, amino acids, and fatty acids. I always wondered why I needed to memorize this complex series of biochemical reactions and similar concepts, and what they could possibly have to do with my future clinical practice. A second-year student recently advised me that questions about the Krebs cycle are often encountered on the United States Medical Licensing Examination (USMLE) Step One examination. But this reason alone is not convincing.

Do practicing physicians need to learn, and subsequently forget, arcane chemical reactions, the molecular structure of—for example—morphine, the branches of the brachial plexus, the side chains on the benzene ring that differentiate toluene from phenol, and the characteristics of liver cells under the microscope? Just ask me if I could describe the Krebs cycle today. I can't. Certainly physicians in very narrow specialties may need to know one or another of these items. But for most of us, they have seemed to be simply obstacles to overcome on the way to treating real live patients.

However, there is another way to look at the study of what might seem to be unnecessary basic science concepts. I recently read (cannot recall where) of a retired professional football player, complaining to his physician of various aches and pains attributed to past injuries. He said to the doctor: "Maybe it would have been better if I had never played football." The doctor replied, "I don't agree. That experience is what made you the man you are."

There is an analogy here. All the basic science experience a medical student endures makes for a better physician in the end. The basic science courses teach mental discipline. They help us learn the language of medicine; one admittedly unscientific source estimates that students learn 15,000 new words over their 4 years of medical school [2]. Consider the current breakthrough in genomic medicine—what was considered "an aspirational term" a decade ago—with the identification of drug response biomarkers and next-generation sequencing in cancer pharmacogenomics [3]. Could a mid-career physician today begin to understand these advances without some grounding in genetics and pharmacology? Where would this physician be if medical school lacked the basic sciences and consisted of 2–3 years of learning diagnostic algorithms and therapeutic guidelines of the day?

What about Spiro's "would-be tailors" analogy? I am certain that before these aspiring tradesmen cut the fabric for their first suits, they had a long apprenticeship in which any theory they had learned was translated into practical skills. It is similar in medicine, as fledgling physicians complete the clinical years of medical school plus residency training to integrate what they have learned in the basic sciences with clinical knowledge and experience.

If you had told me five decades ago that late in my professional life I would be a cheerleader for basic science education, I would have been skeptical. Perhaps it is really true that we get wiser as we grow older.

1. Kravetz RE. Medical humanism: aphorisms from the bedside teachings and writings of Howard M. Spiro. J Fam Pract. 2008;57(10A)Suppo:S27.
2. Joh JW. Quora. Available at: http://www.quora.com/Medical-Students/Approximately-how-many-terms-does-a-1st-or-2nd-year-medical-student-memorize-per-weekday.
3. McCarthy JJ et al. Genomic medicine: a decade of successes, challenges, and opportunities. Sci Transl Med. 2013;5:189.

Chapter 8
Medical Experience, Knowledge, and Wisdom

Medical experience, knowledge, and wisdom: the three, of course, are quite different.

Experience is what we have encountered so far in our lives: what we have seen, heard, felt, loved, hated, endured, and survived. Experience was the data source of the earliest healers, which was balanced, and often in conflict, with ritual and appeals to the supernatural. Today, experience is not as venerated as in the past, surrendering the field to the epidemiologists, statisticians, biometricians, and others who rarely see sick persons outside their own immediate families.

In medicine, we have also become aware that our thinking can be shaped by "outlier" experiences. Osler once remarked that we are "constantly misled by the ease with which our minds fall into the rut of one or two experiences" [1]. As an intern in the Public Health Service Hospital in Norfolk, Virginia, I had the bad fortune to encounter three patients in 1 year with dissecting aortic aneurysms. Based on experience, it seemed to me at the time that aortic dissection was a common disease. It would be several decades before I found my next case.

Which leaves us with knowledge and wisdom. Sir William Osler was fond of repeating to medical audiences the following poem by English poet William Cowper (1731–1800) [2]:

Knowledge and wisdom, far from being one,
Have oft-times no connexion. Knowledge dwells
In heads replete with thoughts of other men;
Wisdom in minds attentive to their own.
Knowledge is proud that he has learned so much;
Wisdom is humble that he knows no more.

© Springer Science+Business Media New York 2015
R.B. Taylor, *On the Shoulders of Medicine's Giants*,
DOI 10.1007/978-1-4939-1335-0_8

Knowledge is what we know, even if some of it will prove to be quite wrong—like the belief that miasma, "bad air," caused cholera, a notion that persisted until the mid-nineteenth century, when John Snow recognized the contagious nature of the disease and famously removed the handle of the Broad Street pump [3]. Today's knowledge, ideally based on the modern scientific method, is less likely than in the past to be proven spectacularly misguided.

What about medical wisdom? A few years ago, I wrote a book titled *Medical Wisdom and Doctoring: The Art of 21st Century Medicine*. Because writing a book with such a title was a presumptuous and risky endeavor, I gave a lot of thought to the definition of *medical wisdom*. In the end, I concluded that medical wisdom is "the capacity to understand and practice medicine in a common-sense manner that is scientifically sound, sensitive to patient needs, ethically grounded, and professionally satisfying" [4]. I believe that most, if not all, of the physician-giants discussed in this book meet these criteria. Many would not have been listed as "Top Doctors" in local Sunday Supplements of their times, and some—like Semmelweis and Halsted—suffered personal afflictions. But all gave us insightful words that enrich our lives.

And so this chapter is about the inquiring thoughts, the observations, the hands-on findings, the research design, the data and its analysis, the conclusions, the promulgation of findings, the wise and selective integration of these "truths" into everyday practice, and the occasional missteps along the way.

1. Osler, W. Teacher and student. In: Aequanimitas, with other addresses, 3rd edition. Philadelphia: Blakiston; 1932.
2. Cowper W. The task. Book VI, Winter walk at noon, 1785;I,88.
3. Competing theories of cholera. Available at: http://www.ph.ucla.edu/epi/snow/choleratheories.html.
4. Taylor RB. Medical wisdom and doctoring: the art of 21st century medicine. New York: Springer; 2010, page 6.

The Roots of Scientific Medicine

> To be matter of scientific knowledge a truth must be demonstrated by deduction from other truths.
>
> Greek philosopher Aristotle (384–322 BCE) [1]

The origins of our current imperative that clinical decisions be grounded in knowledge derived by recognized scientific methods may lie with the thoughts of Aristotle (see Fig. 8.1) and others, such as Hippocrates, who lived during the Golden Age of Greece. Hippocrates held, "We must turn to nature itself, to the observations of the body in health and disease, to learn the truth" [2]. In other words, we should seek observable facts and then use our reason to discover "truth," which today we might call *conclusions*.

Leonardo da Vinci (1452–1519) stated, "And although nature commences with reason and ends in experience, it is necessary for us to do the opposite, that is to commence … with experience and from this to proceed to investigate the reason" [3].

Fig. 8.1 Portrait of Aristotle, student of Plato and teacher of Alexander the Great. http://www.springerimages.com/Images/Medicine AndPublicHealth/5-10.1186_1748-7161-4-6-5

Then in the seventeenth century, Sir Francis Bacon (1561–1626) advocated that scientists "analyze experience and take it to pieces, and by a due process of exclusion and rejection lead to an inevitable conclusion" [4]. The scientific method was evolving and earning respect.

We twenty-first-century physicians have become increasingly committed to scientifically grounded medical practice. The Holy Grail has become the large randomized controlled trial that yields a p value less than 0.05. Of course, we who practiced medicine in the later decades of the twentieth century always believed that our clinical care was (ideally) based on findings reported in the medical literature. Then came *evidence-based medicine* (EBM), a term reportedly first used by investigators at McMaster University in Hamilton, Ontario, Canada, and defined at that time as "a systematic approach to analyze published research as the basis of clinical decision making" [5].

Subsequently we came to recognize that, just as all steaks are not tenderloin, not all evidence is first rate. The US Department of Health and Human Services' National Guidelines Clearinghouse has established an evidence hierarchy ranging from (I) evidence from a single randomized controlled trial or meta-analysis of such trials to (IV) evidence from expert committee reports or opinions and/of clinical evidence of respected authorities [6]. Opinion has yielded to discovery.

Today EBM is widely embraced by physicians and the community of science. We have come a long way in only two and a half millennia.

1. Aristotle. Nicomachean Ethics. Vi.6 (trans. by H Rackham). Cambridge, MA, Harvard University Press; London, William Heinemann Ltd. 1934. Also see: http://www.perseus.tufts.edu/hopper/text?doc=Perseus%3Atext%3A1999.01.0054%3Abekker+page=1140b.
2. Hippocrates. Cited in: Robertson JG. Robertson's words for a modern age. A cross reference of Latin and Greek combining elements. New York: Senior Scribe Publications; 1991, page 183.
3. da Vinci L. In: Manuscript E, in the notebooks of Leonardo da Vinci, Vol I, Chap 19 (trans. by E MacCurdy). Library of the *Institut de France*; 1939.
4. Bacon, F. The great instauration, the plan of the work, 3 vols, Basil Montague, ed. and trans. Philadelphia: Parry & MacMillan; 1854, vol 3, page 333.
5. Claridge JA et al. History and development of evidence-based medicine. World J Surg. 2005;29:547.
6. National Guideline Clearinghouse Guideline Synthesis Template: Available at: http://www.guideline.gov/syntheses/template.aspx.

About Observation and Discovery

I am not accustomed to saying anything with certainty after only one or two observations.

Flemish anatomist Andreas Vesalius (1514–1564) [1]

We recall Vesalius as the anatomist who compiled the seven-volume *De Humani Corporis Fabrica* (On the Fabric of the Human Body) in 1543, and who is considered the founder of human anatomic study. The theater in which he taught can still be visited at the medical school in Padua, Italy. It is noteworthy that Vesalius lived and worked during the Renaissance, the reawakening from the dark centuries of dogma, religious fervor, and scientific apathy (see Fig. 8.2).

A trailblazer of the scientific method, at least of his day, Vesalius might have followed the quotation above with something like "But I do feel justified in drawing conclusions after scores or even hundreds of similar observations." As an anatomist,

Fig. 8.2 Andreas Vesalius. http://www.springerimages. com/Images/Medicine AndPublicHealth/1-10.1007_ s00134-007-0931-5-1

he might have used as an example the fact that, somehow or other, every muscle in the body has an *origin* and an *insertion*. In today's scientific method, he might even have formulated a null hypothesis and set out to see if data caused it to be supported or rejected.

One observation is an anecdote and two anecdotes are not data. Yet occasionally a single observation can, when presented to a prepared mind, set the stage for a noteworthy discovery. About a century ago, a young US Public Health Service physician named Joseph Goldberger (1874–1929) was tasked to find the cause (and perhaps the cure) of pellagra. The prevailing theory was that it was caused by the "pellagra germ." But then why, asked Goldberger, did inmates in mental hospitals contract the disease and yet the staff members were spared? His conclusion: the diets of the two groups differed. After some rigorous and courageous studies, he proved his theory. Today, thanks to diets with adequate vitamins, pellagra is rarely encountered in the USA [2].

Think of the 1928 discovery by Alexander Fleming who, in a serendipitous episode of poor housekeeping, noticed that a mold colony that had invaded a culture plate inoculated with bacteria inhibited bacterial growth in a halo around the mold colony. Fortunately Fleming did not promptly discard the contaminated culture plate. Instead, his observation and subsequent work on the phenomenon led to the eventual development of the first of the bactericidal antibiotics—penicillin [3].

More recently there is John Eng, an endocrinologist working in the US Veteran's Administration system who, following hints from some obscure and almost forgotten prior reports, observed that venom from the Gila monster lizard could stimulate insulin production. Subsequent work led to the 2005 approval of the drug exenatide (Byetta) [4].

And so, while one or two observations cannot support a conclusion, they can be the springboards to important scientific and medical advances.

1. Vesalius A. Letter on the China root. In: O'Malley CD. Andreas Vesalius of Brussels, 1514–1564. Los Angeles: Univ of Calif Press; 1964.
2. Dr. Joseph Goldberger and the war on pellagra. Available at: http://history.nih.gov/exhibits/goldberger/docs/pellegra_5.htm.
3. Fleming A. On the antibacterial action of cultures of a penicillium, with special reference to their use in the isolation of *B. influenzae*. Brit J Exp Path. 1929;10:226.
4. The discovery and development of Byetta (exenatide). Available at: http://www.multivu.com/assets/53897/documents/53897-Exenatide-History-FINAL-original.pdf.

Try the Experiment

I think your solution is just, but why think? Why not try the experiment? (Letter to Edward Jenner, August 2, 1775)

British surgeon and anatomist John Hunter (1828–1793) [1]

In addition to being one of London's leading surgeons of his day, John Hunter conducted studies on a wide variety of entities, including eels, opossums, lizards, owls, and more.

One of Hunter's students, and later his friend, was Edward Jenner (1749–1823), whose observation of the protection against smallpox afforded by a prior case of cowpox led to the development of the smallpox vaccination, a disruptive innovation that saved countless lives. It would be ideal, for the purposes of this essay, if the 1775 letter cited above referred to cowpox, and was encouraging Jenner to undertake his experiment of inoculating 8-year-old James Phipps with cowpox and later challenging his immunity with smallpox. But that experiment occurred much later,

Fig. 8.3 John Hunter. http://
www.springerimages.com/
Images/Medicine
AndPublicHealth/
1-10.1007_978-1-84882-
342-6_1-2

in 1796. The exhortation to "try the experiment," part of a series of letters of Hunter to Jenner, was related to one of the other topics of their ongoing correspondence. Yet, perhaps Hunter's advice was in Jenner's mind when he pondered what to do next when he learned of the protective effect of a cowpox infection.

Hunter was a strong believer in research, and here I focus on early human experimentation and tell of a few others, like Jenner, who subsequently followed this path. As the story goes, attempting to prove that gonorrhea and syphilis were a single disease, Hunter inoculated the prepuce of a subject with gonorrheal pus; the subject also developed a chancre, because the source of the inoculum—a prostitute—was infected with both diseases. There is even a theory that the subject of the experiment was Hunter, himself, but this tale has been vigorously challenged [2, 3] (see Fig. 8.3).

Although the legend of Hunter's alleged experimental misadventure is in doubt, there are more recent instances of scientific self-experimentation. In 1900 Sir Patrick Manson (1844–1922), studying the theory of transmission of malaria by mosquitos, purposely allowed his son and another volunteer to be bitten by malarious mosquitos. Both developed malaria 2 weeks later and were treated with quinine [4].

A few years later Joseph Goldberger, as described above, was attempting to convince the world that pellagra was related to diet and was not infectious. To prove that there was no "pellagra germ," he, his assistant, and his wife conducted "filth parties," in which they swallowed scabs from pellagra patients, swabbed noses of affected patients and then swabbed their own noses, and even injected themselves with blood from "infected" individuals. None of the experimental teams developed pellagra [5].

These events raise some important ethical (and even common sense) issues, but they show the commitment of physician investigators to advancing medical knowledge.

1. Letters from the past from John Hunter to Edward Jenner. Ann R Coll Surg Engl. 1975;57:215.
2. Wright DJM. John Hunter and venereal disease. Ann Roy Coll Surg Engl. 1981;63:198.
3. Gladstein J. Hunter's chancre: did the surgeon give himself syphilis? 2005;41:128. Available at: http://history.nih.gov/exhibits/goldberger/index.html.
4. Manson P. Experimental proof of the mosquito-malaria theory. BMJ 1900;2:949.
5. Joseph Goldberger and his war on pellagra. Office of NIH History. Available at: http://history.nih.gov/exhibits/goldberger/.

On Caution with Hypothetical Statements

The advantages which have been derived from the caution with which hypothetical statements are admitted are in no instance more obvious than those sciences which more particularly belong to the healing art.

English surgeon and naturalist James Parkinson (1755–1824) [1]

The above is taken from the preface to the highly accurate description of the shaking palsy, or "paralysis agitans," which would later be called Parkinson disease (see Fig. 8.4). Translated from the somewhat turgid prose of the early nineteenth century, Parkinson urges great care in drawing scientific conclusions, especially when human health may be affected.

Fig. 8.4 James Parkinson (Public domain). http://commons.wikimedia. org/wiki/File: James_Parkinson.jpg

History is replete with examples of scientific theories that seemed plausible in their day and were later proved wrong:

- The four bodily humors: black bile, yellow bile, phlegm, and blood
- The geocentric universe, with the earth at the center of the universe
- Spontaneous generation: life springing from inanimate matter
- The flat earth theory and the hollow earth theory

I am not sure that great harm resulted from any of these misbeliefs. But the outcome can be different when the theories affect health decisions. Consider, for instance, the Andrew Wakefield hypothesis linking the administration of measles, mumps, and rubella (MMR) vaccine to the development of autism and bowel disease, reported in a paper published in *The Lancet* in 1998, and subsequently retracted in 2010 [2, 3]. The Wakefield study involved a grand total of 12 children whose parents described some loss of previously acquired developmental skills and who were also found to have some sort of "intestinal abnormalities." The authors report: "Onset of behavioural symptoms was associated, by the parents, with measles, mumps, and rubella vaccination in eight of the 12 children, with measles infection in one child, and otitis media in another" [2].

The upshot of this study, later described as "fraudulent," was a stunning loss of confidence in the safety of the MMR vaccine and subsequent decline in the number of children who received MMR vaccine [4].

Taken in context of a tiny (a dozen) uncontrolled subject cohort, the reliance on parents' stories, and the truly soft data, it is a wonder that the Wakefield report had a dozen (although perhaps witless) coauthors, that the manuscript was not rejected out of hand by the person who opened the envelope at the journal, and that it survived the peer review process of a respected medical journal and found its way into print. All of this reinforces Parkinson's admonition to view hypothetical statements with caution when they pertain to the healing art—and to real people.

1. Parkinson J. An essay on the shaking palsy. Preface. London: Whittingham and Rowland; 1817.
2. Wakefield AJ et al. Ileal-lymphoid-nodular hyperplasia, non-specific colitis, and pervasive developmental disorder in children. The Lancet. 1998;351:637 [RETRACTED].
3. News: Lancet retracts Wakefield's MMR paper. BMJ. 2010;342:c696.
4. Godlee F, et al. Wakefield's article linking MMR vaccine and autism was fraudulent. BMJ. 2010;342:c7452.

Chance and the Prepared Mind

Chance only favors the prepared mind.

French microbiologist and chemist Louis Pasteur (1822–1985) [1]

Louis Pasteur's work cast a wide net, including beer and wine, silkworms, polarized light, and milk spoilage. His name is linked with the germ theory of disease, vaccination against anthrax, and of course, what we today call Pasteurization. But one of his greatest discoveries came as he was studying chicken cholera, caused by an organism in the genus we now call *Pasteurella*. Almost by chance, Pasteur discovered that the organisms lost virulence when successively re-cultured, and that the weaker organisms could induce resistance to the full-strength bacteria [2] (see Fig. 8.5).

How is it that several people can see the same phenomenon, but only one recognizes its significance? I am reminded of the quip attributed to Sir Winston Churchill:

Fig. 8.5 Louis Pasteur.
http://www.springerimages.
com/Images/Medicine
AndPublicHealth/
1-10.1007_978-1-4614-
4845-7_7-1

"Men occasionally stumble over the truth, but most of them pick themselves up and hurry off as if nothing ever happened" [3]. But there are exceptions to Churchill's adage. In 1941, Australian ophthalmologist N. McAlister Gregg connected the dots and first reported the connection between maternal rubella in early pregnancy and the occurrence of cataracts in the children born of these mothers [4].

In 1982 Australian physician-scientists Barry Marshall and Robin Warren found the organism *H. pylori* in the gastric mucosa of peptic ulcer patients and hypothesized that the microorganism, and not stress, as was believed at the time, was the cause of peptic ulcer disease (PUD). To prove his point, Warren even drank a culture of *H. pylori* and subsequently developed PUD manifestations, which he cured with antibiotics [5].

And in 1983, American physician Bruce Tempest, chief of medicine at the Indian Medical Center in Gallup, New Mexico, noted a small cluster of illnesses with a high mortality. Sensing something unusual, he set in motion a series of investigations leading to the recognition of a Hantavirus infection, a disease seldom seen outside of Asia at that time [6].

There are many examples of physicians and scientists whose prepared minds recognized the significance of chance observations. Now if only we could recognize how to "prepare minds" to see meaning in the unexpected, and if we could somehow teach this to our students.

1. Pasteur L. Inaugural lecture. University of Lille, December 7, 1854. In: Vallery-Radot R. The life of Pasteur. Devonshire RL, trans. Garden City NY: Garden City Publishing; 1923, page 76.
2. Louis Pasteur. In: Encyclopedia Brittanica. Available at: http://www.britannica.com/EBchecked/topic/445964/Louis-Pasteur/281417/Vaccine-development.
3. Quote investigator. Available at: http://quoteinvestigator.com/2012/05/26/stumble-over-truth/.
4. Gregg N McA. Congenital cataract following German measles in the mother. Trans Ophthalmol Soc Aust. 1941;335:46.
5. Marshall BJ, Warren JR. Unidentified curved bacilli in the stomach of patients with gastritis and peptic ulceration. The Lancet. 1984;323:1311.
6. Grady D. Death at the Corners. Discover. December, 1963. Available at: http://discovermagazine.com/1993/dec/deathatthecorner320-.UlG7WBbXJ_a.

About Outside Influences on Medicine

The truth is that medicine, professedly founded on observation, is as sensitive to outside influences, political, religious, philosophical, imaginative, as is the barometer to the changes in atmospheric density.

American physician and author Oliver Wendell Holmes, Sr. (1809–1894) [1]

Oliver Wendell Holmes, like William Carlos Williams, was both physician and author (see Fig. 8.6). In the latter role he became famous for his "Breakfast Table" series of essays. Holmes mentions observation of disease processes—the foundation of Hippocratic medicine. In fact, it would be reasonable to assert that the Ionian philosophy of observation leading to rational conclusions set the stage for today's science-based medicine. But much happened in the meantime, and we have no shortage of influences on medicine to report over the intervening years [2].

The laws of early Rome prohibited human dissection; as an alternative, Galen (129–200) dissected animals, including pigs and monkeys, and presumed that they

Fig. 8.6 Oliver Wendell Holmes, Sr. http://www.springerimages.com/Images/MedicineAndPublicHealth/1-10.1007_s00129-003-1422-6-1

shared like anatomy, thus introducing muddled medical beliefs that survived for a millennium [3]. In medieval times, Church teachings held that illness was God's punishment for sins, and some even believed that the practice of medicine was not an appropriate activity for Christians [4]. For centuries, the Catholic Church has influenced the delivery of reproductive health care services in many countries [5]. And on July 22, 1934, the German Government excluded Jews from National Health Insurance [6]. As I write this, the ironically titled US Affordable Care Act is causing thousands of Americans to lose long-standing health insurance and others to experience increased insurance costs.

Today's influences on medicine come less from religious, philosophical, or imaginative sources than from the political forces that affect so many facets of our daily life. We are sometimes told what we can and cannot advise patients about their reproductive options. We are instructed how much we can and cannot charge patients for services we render. We might be prosecuted if we provide *professional courtesy* to colleagues; that is, charge them less than we charge other patients. Restaurants may "refuse service to anyone," but we have lost that right. Decisions about who can have access to scarce resources are being taken out of the hands of physicians. The list goes on.

Holmes shed light on the outside influences on medicine. Today, more than ever, forces beyond our cozy medical world will affect our patients' welfare. We physicians must be involved in shaping the changes that will inevitably occur. If we do not, someone else will make the influential decisions.

1. Holmes OW, Sr. Medical essays: current and counter-currents in medical science. Cambridge MA: Riverside Press; 1861.
2. Nuland SB. The biography of medicine. New York: Random House; 1988.
3. Nutton V. The unknown Galen. London: Institute of Classical Studies, School of Advanced Study, University of London; 2002, page 89.
4. Wallis F. Medieval medicine: a reader. Toronto: University of Toronto Press; 2010.
5. Markwell HJ et al. Bioethics for clinicians: Catholic bioethics. CMAJ. 2001;165:189.
6. Holocaust timeline. The history place. Available at: http://www.historyplace.com/worldwar2/holocaust/timeline.html.

Medical History as the History of Humanity

> The history of medicine is, in fact, the history of humanity itself, with its ups and downs, its brave aspirations after truth and finality, its pathetic failures. The subject may be treated variously as a pageant, an array of books, a procession of characters, a succession of theories, an exposition of human ineptitudes, or as the very bone and marrow of cultural history.
>
> American medical historian Fielding H. Garrison (1870–1935) [1]

The book you are reading is, at its core, a medical history book. The seminal work in this field was written by Garrison a century ago, and it stands today as a triumph of medical scholarship. As an intern in 1961, I bought a copy of the fourth edition and read it cover to cover; I still have this book with all my underlines and margin notes. I think it sparked my interest in the history of our noble profession.

From prehistoric healing rituals to today's telemedicine, advanced imaging, and robotic-assisted surgery, no profession—not the clergy, not the law, certainly not business—is as intertwined in the fabric of humankind as is medicine. Humankind, of course, is not always rational and wise.

In this book, I present the thoughts of many of best, and their aspirations for truth such as the hunts for cures of malaria, syphilis, and leukemia. We have also seen medicine's failures: the perpetuation of some of Galen's misbeliefs for centuries and the fervent opposition to Semmelweis' theory of the origin of childbed fever.

Part of medicine's colorful history not featured in this book—because its practitioners were not giants of intellect and integrity—involves the pretenders, the charlatans, and the quacks. For example, the "cereal doctors" had some curious methods. John H. Kellogg (1852–1943) advocated creating a healthy bowel using an enema machine that delivered 15 gallons of water in a few seconds. C.W. Post held that appendicitis could be aborted by consuming large quantities of, you guessed it, *Grape-Nuts* [2].

In the early twentieth century, Doctor John R. Brinkley (1885–1942) offered that, for a mere $750, he could cure male impotence and fertility by the surgical implantation of goat glands. At about this same time, Doctor Albert Abrams, once vice president of the California Medical Association, introduced the Electronic

Fig. 8.7 Dr. Albert Abrams (Public domain). http://commons.wikimedia.org/wiki/File:Dr._Albert_Abrams.jpg

Reactions of Abrams (ARA), the force behind his invention, the Dynomizer (see Fig. 8.7). This was a machine that could diagnose the patient's ailment from a drop of blood or even a sample of handwriting; the patient could then be offered a cure using vibratory waves provided by the Oscilloclast. By 1921, some 3,500 clinicians were using the ARA in their practices [2, 3]. In the late twentieth century, how many desperate patients received laetrile therapy for their cancers, some even after its toxicity and ineffectiveness had been demonstrated?

All the above reflect the very human tendency to seek relief from suffering, even when an idea seems just a little fanciful.

1. Garrison FH. Preface: An introduction to the history of medicine. Philadelphia: Saunders; 1913.
2. Taylor RB. White coat tales: medicine's heroes, heritage, and misadventures. New York: Springer; 2007, pages 222–230.
3. Fishbein, M. The medical follies: an analysis of the foibles of some healing cults, including osteopathy, homeopathy, chiropractic, and the electronic reactions of Abrams, with essays on the anti-vivisectionists, health legislation, physical culture, birth control, and rejuvenation, New York: Boni & Liveright; 1925.

Not Quite Everything Discovered Yet

> On entering the practice of medicine (circa 1890) I was under the impression that the knowledge of disease was so nearly complete that there was absolutely nothing to be done but to apply that knowledge to my patients .... I made little progress, and after several years it gradually dawned on me that the field of medicine had not, perhaps, been so fully explored as I had thought, and I resolved that I would begin to observe for myself.
>
> Scottish physician Sir James Mackenzie (1853–1925) [1]

We remember Mackenzie for observations he made on his patients over his years of practice. He developed a device allowing him simultaneous measurement of arterial and venous pulses, and he pioneered the use of digitalis in patients with arrhythmias. The discovery of premature ventricular contractions is attributed to him [2] (see Fig. 8.8). With Mackenzie as a model, let us look at three other physicians who made personal observations of their patients over the years.

Fig. 8.8 Sir James Mackenzie. http://www. springerimages.com/search. aspx?caption=James Mackenzie

In writing the words quoted above, one could imagine that Mackenzie was thinking of William Beaumont (1785–1853), a US Army surgeon who advanced medical science by recording longitudinal observations on gastric physiology. Beaumont's "series-of-one" subject was Alexis St. Martin, a fur company employee who suffered an accidental shotgun wound to the upper abdomen in 1822. The wound resulted in a permanent fistula in the stomach, and for a decade, Beaumont conducted studies on gastric fluids and factors affecting digestion through the window into his patient's stomach. In 1833, Beaumont published the results of his observations [3].

In the 1930s country doctor William (Will) Pickles (1885–1969) in England traced an epidemic of catarrhal jaundice, aka viral hepatitis A, in 250 people to a single source child, and identified the disease's long incubation period of 26–35 days [4].

Curtis G. Hames (1920–2005), a family doctor in rural Evans County, Georgia, noted in the 1950s that there seemed to be racial differences in the incidence of hypertension and coronary artery disease. In 1959 he enrolled most of the county residents in what was to become a long-term research program. Participants received a physical exam, laboratory testing, and an electrocardiogram; and, what later proved invaluable, he banked frozen serum to be used in later studies. The studies chiefly concerned determinants of cardiovascular disease and, with National Institutes of Health (NIH) support, research involving members of the cohort continued for 37 years, until 1995 [5, 6].

If you or I should ever think that there are no discovery opportunities left, consider the ill-fated words attributed to Charles H. Duell, US Patent Office Commissioner, in 1899: "Everything that can be invented has been invented."

1. Mackenzie J. On the teaching of clinical practice. Br Med J. 1914;1:17.
2. Hamish JC et al. Profiles in cardiology: Sir James Mackenzie. Heart views. 2001:2.
3. Beaumont W. Experiments and observations on the gastric juice and the physiology of digestion. Plattsburgh NY: F. P. Allen; 1833.
4. Pickles W. Epidemiology in country practice. Proc Roy Soc Med. 1935;28:1337.
5. Curtis G. Hames, Sr, MD—the scholar. Medical College of Georgia. Available at: http://www.georgiahealth.edu/medicine/fmfacdev/hamesthescholar.html.
6. Curtis G. Hames, Sr, MD—the physician. Medical College of Georgia. Available at: http://www.georgiahealth.edu/medicine/fmfacdev/hamesthephysician.html.

Not Always by Accumulation of Data

> Great discoveries which give a new direction to currents of thought and research are not, as
> a rule, gained by accumulation of vast quantities of figures and statistics.
>
> American physician and scientist Theobald Smith (1859–1934) [1]

Smith investigated a number of diseases affecting humans and animals. He differentiated between human and bovine tuberculosis, described the phenomenon of anaphylaxis, and, especially pertinent for this essay, considered the possibility that malaria may be spread by mosquitos [2, 3].

Of course, many important discoveries have followed extensive trial and error involving hundreds of possibilities. In 1910, arsphenamine (Salvarsan), discovered in the laboratory of Paul Ehrlich and marketed by Hoechst AG for the treatment of syphilis, was originally dubbed "606." Why? Because it was number six in the sixth set of compounds tested by Ehrlich [4].

Later, in the years leading up to World War II, German scientists developed sulfonamides, the world's first true antibacterial drug class, by testing and rejecting hundreds of coal tar dyes and later by formulating and discarding scores of variations on the original compound discovered [5].

Recently the drug imatinib (Gleevec), used to treat chronic myelogenous leukemia, was developed as "an initial lead compound was identified by the time-consuming process of random screening, that is, the testing of large compound libraries for inhibition of protein kinases in vitro" [6].

In contrast, and in support of Smith's thesis, I offer the tale of Sir Ronald Ross (1857–1932), who, in 1897, discovered that malaria is spread by the mosquito by finding the parasite in blood from a mosquito that had been allowed to feed on the blood of a known malaria patient. This discovery and Ross's subsequent work on malaria prevention have saved countless lives [7]. Ross's discovery seems to have come from clinical acumen, courage, and persistence, rather than by assembling a mountain of data.

Here is a poem by English poet John Masefield (1878–1967), who was a close friend of Ross. The poem was published In *The Times* on August 20, 1957, to commemorate the discovery by Ross of the role of the mosquito in transmitting malaria 60 years before [8].

A Moment on the August day, 1897

Once, on this August Day, an exiled man
Striving to read the hieroglyphics spelled
By changing speckles upon glass, beheld
A secret hidden since the world began.

The limitless instant of joy laid bare
A force to alter life, a key, a clue
Delaying Death and all its retinue;
Light showing as all answer to one prayer

1. Smith T. Scholarship in medicine. Boston Med Surg J. 1915;172:121.
2. Dolman CE et al. Suppressing the diseases of animals and man: Theobald Smith, Microbiologist. Boston: Boston Medical Library; 2003.
3. Nuttall GHF. Theobald Smith: 1859–1934. Obituary notices of the Fellows of the Royal Society. 1935;1:514.
4. Gibaud S et al. Arsenic-based drugs. In: Fowler's solution to modern anticancer therapy. Topics Organometallic Chem. 2010;32:1.
5. Hager T. The demon under the microscope: from battlefield hospitals to Nazi labs, one doctor's heroic search for the world's first miracle drug. Old Saybrook, CT: Tantor Media; 2006.
6. Drucker BJ et al. Lessons learned from the development of an Abl tyrosine kinase inhibitor for chronic myelogenous leukemia. J Clin Invest. 2000;105:3.
7. Biography of Sir Ronald Ross. London School of Hygiene & Tropical Medicine. Available at: http://www.lshtm.ac.uk/library/archives/ross/biography/.
8. Masefield J. A moment on the August day, 1897. The Times. August 20, 1957.

Believing in Your Work

If I didn't believe that the answer could be found, I would not be working on it.

American physician and scientist Florence Rena Sabin (1871–1953) [1]

Dr. Florence Sabin achieved an impressive record of "firsts" during her long career in medicine and science: She served her internship in 1900–1901 under Sir William Osler, and in 1902 she was the first woman appointed to the faculty of the Johns Hopkins School of Medicine in Baltimore. In 1924, she became the first woman president of the American Academy of Anatomists. Then, a year later, she became the first woman to be a full member of the Rockefeller Institute, and was the first woman to be elected to membership in the National Academy of Sciences [2].

Fig. 8.9 Rena Sabin (Public domain). http://commons.
wikimedia.org/w/index.php?s
earch=Rena+Sabin&title=Sp
ecial%3ASearch&go=Go&us
elang=en

Her work included study of the brain, the lymphatic system, blood cells, and connective tissue. While at the Rockefeller Institute, she studied the pathology of tuberculosis, and in 1945, she received the Trudeau Medal presented by the National Tuberculosis Association for her "extensive studies of the physiologic activities of the chemical fractions of the tubercle bacillus" [3] (see Fig. 8.9).

In 1929, when accepting one of her many awards, she commented: "I hope my studies may be an encouragement to other women, especially to young women, to devote their lives to the larger interests of the mind. It matters little whether men or women have the more brains; all we women need to do to exert our proper influence is just to use all the brains we have" [4].

Returning to the quotation above about believing in your work, we find a fierce determination to find answers to questions, and also to use time wisely; that is; as she advised young women, "to use all the brains we have." This serious attention to purpose, in fact, characterizes so many of the successful scientists described in this book, and especially in this chapter: Vesalius, Hunter, Pasteur, Mackenzie and Szent-Györgyi, and others.

If Ehrlich, in 1910, had given up after testing, let's say, 600 compounds, he would have never found Compound 606, Salvarsan. If there had been an early declaration of failure, we would have no Gleevec. Perhaps Sabin's quotation—*"If I didn't believe that the answer could be found, I would not be working on it"*—should be displayed on the wall in every research laboratory.

1. Sabin FR. Quoted in: Karnes FA et al. Young women of achievement: a resource for girls in science, math, and technology. Amherst NY: Prometheus Books; 2002.
2. McMaster PD et al. Biographical memoirs: Florence Rena Sabin. The National Academies Press. Available at: http://www.nap.edu/readingroom.php?book=biomems&page=fsabin.html.
3. Changing the face of medicine: Dr. Florence Rena Sabin. Available at: http://www.nlm.nih.gov/changingthefaceofmedicine/physicians/biography_283.html.
4. Sabin FR. Acceptance remarks: Pictorial Review achievement award, 1929. Available at: Profiles in Science. National Library of Medicine. Available at: http://profiles.nlm.nih.gov/ps/retrieve/Narrative/RR/p-nid/89.

What Nobody Else Has Thought

> Discovery consists of seeing what everybody has seen and thinking what nobody has thought.
>
> Hungarian physiologist Albert Szent-Györgyi (1893–1986) [1]

Szent-Györgyi isolated ascorbic acid, aka vitamin C, and for this achievement he received the 1937 Nobel Prize in Physiology or Medicine (see Fig. 8.10). His work also helped elucidate the reactions that constitute the Krebs citric acid cycle, discussed in Chap. 7 [2]. These are both noteworthy scientific discoveries.

There are, of course, serendipitous discoveries. Christopher Columbus stumbling upon America on his way to India comes to mind as an example. Ambroise Paré's switch from wound cautery and boiling oil to a soothing ointment as a cost-cutting measure is another. But many more discoveries involve a "Eureka" moment in which the observer senses some connection, relationship, or meaning not previously appreciated.

Fig. 8.10 Albert Szent-Györgyi. http://www.springerimages.com/Images/Physics/1-10.1007_978-3-642-33430-6_1-1

In many cases we know just who had the epiphany and when: In 1775 William Withering (1741–1799), in England, hit on the idea that one ingredient—digitalis, derived from the foxglove plant—in a brew employed by an elderly folk herbalist could help relieve "dropsy" (heart failure).

Earlier in this chapter I told the story of Edward Jenner and his recognition in 1796 of the potential value of "vaccination." Here is one example closer to home: While working at the 3M Company in 1974, Arthur Fry was faced with a glue that didn't stick well; instead of discarding the compound as faulty, he sensed its potential and the Post-It note was born.

In other instances, the individual who had the insightful moment and just when it occurred remain less well known: The stimulant methylphenidate (Ritalin) is used to treat attention-deficit hyperactivity disorder. How did anyone ever think that giving a stimulant to hyperactive children might be beneficial? Botulinum toxin (Botox) was first used for therapeutic purposes in the late 1960s. What prompted someone to speculate that this lethal toxin could have clinical uses? [3].

In the early 1980s physicians encountered increasing numbers of patients with various uncommon diseases: *Pneumocystis carinii* pneumonia, Kaposi sarcoma, and unexplained widespread lymphadenopathy. Then it was found that these diseases tended to occur in four cohorts: homosexuals, hemophiliacs, Haitians, and heroin users. Eventually someone recognized that, in fact, this was a single disease—acquired immunodeficiency syndrome or AIDS—even before the viral cause was discovered in 1983 [4].

All this makes me wonder what we are all seeing today that eventually, when viewed through the eyes of someone who has Pasteur's "prepared mind," who thinks what nobody before has thought, will be found to be something unexpected and amazing.

1. Szent-Györgyi A. Quoted in: Good I. The scientist speculates. New York: Basic Books; 1963.
2. Kyle RA et al. Albert Szent-Györgyi—Nobel laureate. Mayo Clinic Proc. 2000:75:722.
3. Erbguth FJ. Historical notes on botulism, *Clostridium botulinum*, botulinum toxin, and the idea of the therapeutic use of the toxin. Movement Disorders. 2004;19:S2.
4. Gallo RC et al. Isolation of human T-cell leukemia virus in acquired immunodeficiency syndrome (AIDS). Science. 1983;220:865.

Chapter 9
Errors, Uncertainty, and Ethical Issues

Much of medicine is about clinical outcomes, and the paths we take to achieve them. Even before seeing our first patients as newly minted physicians, we seek knowledge and "truth" through the study of the works of science's giants who have gone before us: the anatomic discoveries of Vesalius, the microbiology of Pasteur, and the antibiotics that followed Fleming's chance observation that a mold could inhibit bacterial growth. Then we begin seeing patients and do our best to apply what we have learned.

Here is where we learn that the road to diagnosis is often circuitous, with opportunities to go astray. We call these diagnostic errors, when the headache we assumed to be migraine turns out to be tumor, and when the patient with apparent indigestion suffers an aortic dissection. Over the years, medicine's giants have suffered errors, and have probably thought about their meaning. As an example, Robert Koch (1843–1910) discovered the tubercle bacillus and, for this achievement, received the 1905 Nobel Prize in Physiology or Medicine. But Koch's misunderstanding of the differences between human and bovine tuberculosis delayed the discovery that milk could transmit the disease [1, 2] (see Fig. 9.1).

Medical uncertainty is part of daily practice: Is my diagnosis correct? Have I recommended the right therapy? Should I operate now or wait to see what tomorrow brings? No matter how we strive to eliminate dreaded uncertainty, we encounter it again and again. Wellbery, however, sees medical uncertainty as an opportunity. "In medicine—a field where the physical body registers palpable outcomes—certainty about diagnosis, therapy, and prognosis is a logically desirable goal. In the arts, by contrast, uncertainty and ambiguity are often embraced because they create opportunities, moving the perceiver beyond the obvious into a realm where values,

© Springer Science+Business Media New York 2015
R.B. Taylor, *On the Shoulders of Medicine's Giants*,
DOI 10.1007/978-1-4939-1335-0_9

Fig. 9.1 Robert Koch. http://
www.springerimages.com/
Images/LifeSciences/
1-10.1007_978-1-4419-1108-
7_3-6

meanings, and priorities are weighed and adjudicated. I would like to suggest that this exploratory potential applies to medical practice. While clearly, clinical medicine seeks assurances, it is also an 'art', whose practitioners acknowledge, and even thrive upon, the messiness and complexity it targets. There is, then, a positive role for medical uncertainty that can serve as a counterforce to our unexamined quest for definitive answers" [3].

Also part of daily practice are ethical issues, as we balance our treasured values of beneficence, non-maleficence, confidentiality, patient autonomy, and more. All is well until a specific case presents dissonance between two perfectly legitimate ethical values, as when a patient with an illness that could cause fainting asks that his family not be told. This pits the patient's right to confidentiality against the issue of non-maleficence in regard to the family members who may be passengers in the patient's car or perhaps the driver of an approaching vehicle.

This chapter examines some of the gray areas in medicine and what our giants had to say about them.

1. Cobbet L. The relation between animal and human tuberculosis. In: The causes of tuberculosis. Cambridge, MA: Cambridge University Press; 1917.
2. Palmer MV et al. Bovine tuberculosis and the establishment of an eradication program in the United States: role of veterinarians. Vet Med Int. 2011; Article ID 816345.
3. Wellbery C. The value of medical uncertainty? The Lancet. 2010;375:1686.

First, Do No Harm

The physician must … make a habit of two things: to do good or at least to do no harm.

Hippocrates (ca. 460–377 BCE) [1]

Often expressed in its Latin version—*Primum non nocere*—the admonition to do no harm is deeply imbedded in the lore of Western medicine. No chapter discussing medical errors would be complete without some reference to the venerated saying.

Although often attributed to Hippocrates, "first, do no harm," as expressed in various languages, has a rich history. One may argue that the earliest use of the caution was by the Yellow Emperor Huang Ti (ca. 2600 BCE), who wrote: "The most important requirement of the art of healing is that no mistakes or neglect occurs" [2]. Smith attributes the specific expression, with its Latin version, to Thomas Sydenham (1624–1689), whom we remember for his classic descriptions of disease

Fig. 9.2 Teenage girl with weight loss caused by anorexia nervosa. http://commons.wikimedia.org/wiki/Category:Anorexia_nervosa#mediaviewer/File:Anorexia_case_1900.jpg

[3]. Herranz tells that the maxim to "first, do no harm" entered the fabric of British and American medicine in 1849 when it was used in a book by Worthington Hooker titled *Physician and Patient* [4]. Today, I would wager that, whether in English or in Latin, there is not a US medical student who has survived to the clinical years who is not familiar with this adage.

Doing harm can occur in degrees. We have all heard of the wrong patient receiving medication. Early pregnant uteri have been incorrectly diagnosed as tumors and excised. At the University of Washington Medical Center in June 2000, surgeons operating on Donald Church, age 49, left a 13-in.-long retractor in his abdomen, necessitating a second surgery to remove the "souvenir" [5].

Not all poor outcomes involve such epic misadventures. Sometimes harm to patients falls under the heading of unintended consequences. As an example, O'Dea describes how, in response to the epidemic of childhood obesity, we have advocated dietary control and exercise as sensible preventive measures. She holds that: "The moderate and sensible dietary guideline of the late 1970s was taken up by the target audience who required it least—young women, who adhered to the 'control your weight' most vehemently," leading to an "exponential rise in eating disorders" [6] (see Fig. 9.2).

All physicians commit errors and cause harm. Our daily work involves making many decisions, often with incomplete data, and with the stark realization that we cannot predict the future. Unfavorable outcomes happen. We can only hope that our errors will be "minor" and that little damage results. Hippocrates is reported to have said, "I would give great praise to the physician whose mistakes are small, for perfect accuracy is seldom to be seen" [7]. I do believe that, at the end of one's career, the memorable cases of a physician's lifetime will feature many of the mistakes made.

1. Hippocratic corpus: Epidemics. Book I, Section 11.
2. Huang Ti (Yellow Emperor): Nei Ching Su Wên. Book IV, Section 13.
3. Smith CM. Origin and uses of *primum non nocere*—above all, do no harm. J Clin Pharmacol. 2005;45:371.
4. Hooker W. Physician and patient: a practical view of the mutual duties, relations and interests of the medical profession and the community. New York: Baker and Scribner; 1849.
5. Ten unbelievable medical mistakes. Available at: http://www.oddee.com/item_96576.aspx.
6. O'Dea JA. Prevention of childhood obesity: "First, do no harm." Health Ed Res. 2004;20:259.
7. Hippocrates. On ancient medicine, Book IX.

On Being Very Wrong

I have formerly said that there was but one fever in the world. Be not startled, Gentleman, follow me and I will say that there is but one disease in the world.

American physician and educator Benjamin Rush (1745–1813) [1]

Rush goes on to explain, "The proximate cause of disease is irregular convulsive…action in the (vascular) system affected." In support of his thesis, he asserts "the multiplication of diseases…(is) as repugnant to truth in medicine as polytheism is to religion. The physician who considers every affectation of the different systems of the body…as distinct diseases when they arise from one cause resembles the Indian or African savage who considers water, dew, ice, frost and snow as distinct essences" [1]. Rush was, of course, spectacularly wrong. His pronouncement came eight centuries after Rhazes, in Persia, recognized that the infectious rashes—measles, varicella, and smallpox—were distinct diseases.

A leading physician of his day, Rush was Professor of Medical Theory and Clinical Practice at the University of Pennsylvania Medical School, the founder of Dickinson College in Pennsylvania, and a signer of the Declaration of Independence (see Fig. 9.3). He trained the physicians who attended George Washington at the

Fig. 9.3 Benjamin Rush (Public domain). http://commons.wikimedia.org/wiki/Benjamin_Rush

time of his death [2]. I include his misguided theory here to balance the sagacity of others in this book and to show that even the most venerable physician can espouse a theory that is very mistaken.

There are other examples of scientific wrongheadedness. In 1872, Pierre Pachet, Professor of Physiology at Toulouse University in France, declared, "Louis Pasteur's theory of germs is ridiculous fiction" [3].

Félix Martí-Ibáñez (1911–1972), a Spanish-American physician-writer and Professor of the History of Medicine at the New York Medical College, wrote in 1958: "The profound change that is taking place in the natural history of infections warrants the prophecy that by the year 2000 the diseases caused by bacteria, protozoa, and perhaps viruses will be considered by the medical student as exotic curiosities of mere historical interest, as is the case today with tertiary syphilis, gout, and smallpox" [4]. Then in 1988, a molecular biology professor at University of California, Berkeley, is reported to have described the human immunodeficiency virus (HIV) as "a pussycat" [3].

Off-target predictions are not limited to medicine. Here are a few whoppers from other arenas [5]:

- "The Americans have need of the telephone, but we do not. We have plenty of messenger boys."—*Sir William Preece, chief engineer of the British Post Office, 1876.*
- "Stocks have reached what looks like a permanently high plateau."—*Irving Fisher, Professor of Economics, Yale University, 1929.*
- "I think there is a world market for maybe five computers."—*Thomas Watson, chairman of IBM, 1943.*

Which of our current beliefs/predictions just might prove to be very wrong? Will it be global warming, the theory of evolution through natural selection, or the belief that we can find an effective HIV vaccine?

1. King LS. The medical world of the eighteenth century. Chicago: University of Chicago Press; 1958, pages 223–224.
2. The Benjamin Rush prescription. Providentia: Available at: http://drvitelli.typepad.com/providentia/2012/01/the-benjamin-rush-prescription.html.
3. Frater J. 15 Extremely embarrassing science predictions. Available at: http://listverse.com/2010/12/22/15-extremely-embarrassing-science-predictions/.
4. Martí-Ibáñez F. Men, molds and history. New York: MD Publications: 1958, page 20.
5. Things people said. Bad predictions. Available at: http://www.rinkworks.com/said/predictions.shtml.

About Admitting Errors

> Next to the promulgation of the truth, the best thing I can conceive that a man can do is the
> public recantation of an error.
>
> British surgeon Joseph Lister (1827–1912) [1]

We remember Baron Lister for three things. First of all, he took some of Semmelweis' work to the next level and, beginning in the 1860s, he championed the use of carbolic acid, aka phenol, to sterilize surgical instruments and to reduce bacterial contamination of operative fields [2]. There is the bacterial genus *Listeria*, which includes *Listeria monocytogenes*, the cause of listeriosis (see Fig. 9.4). There is also Listerine mouthwash, initially compounded in 1879, marketed as a surgical antiseptic, and eventually sold as a mouthwash containing more than 20 % alcohol. Listerine is now a staple on drugstore shelves. Dr. Lister, however, was not pleased by the eponymous trademark; he mounted an expensive and unsuccessful campaign to remove his name from the product [3].

Lister's words quoted above, actually referring to an error made in an investigation into the nature of fermentation and not concerning direct patient care,

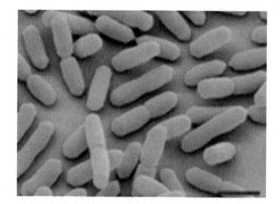

Fig. 9.4 Listeria monocytogenes. http://www.springerimages.com/Images/ Biomedicine/3-27395_0_ Listeria_monocytogenes_ EGD01_BIL140207

nevertheless seem especially relevant today. Currently, in the early twenty-first century, dominant themes in health care include quality assurance, cost-effective care, patient safety, and, pertinent to the topic at hand, transparency as regards errors in medicine.

The issue of facing medical errors—described in Chap. 6 as occurring in 5–15 % of hospitalized patients—was highlighted by the 1999 Institute of Medicine (IOM) report *To Err is Human* [4]. The document emphasized that errors are a "system problem," and nonjudgmental transparency is the first step in what was termed *crossing the quality chasm* and making patients safer.

But when it comes to disclosing medical errors to patients, the decision as to what to do is not always crystal clear. A foreign body left in an abdomen following surgery is an error meriting full disclosure, and remediation. But if a nurse omits a bedtime sedative, and the patient still has a good night's sleep, is this an error reportable to the patient? If the physician misjudges the significance of epigastric distress and some time later the patient is found to have a peptic ulcer or even gastric cancer, is this an error to be reported even if subsequent treatment is successful? When an error occurs, patients want information as to the nature and cause of the misstep, and also an apology [5]. But how many minor missteps can be reported to the patient before there is a loss of confidence?

In Lister's day, care was "physician centered," patients believed that "doctor knows best," physicians were generally held in high regard, and there were no malpractice attorneys hovering in the background. Things have changed, patients are safer, there is greater attention to quality, and we have a recently acquired, epiphanic commitment to view errors as systems problems without first leaping to blame and punish.

1. Godlee RJ. Lord Lister, ed. 2. London: Macmillan; 1918, page 278.
2. Lister J. On the antiseptic principle of surgery. BMJ. 1867;2:245.
3. Dirckx JH. The language of medicine: its evolution, structure and dynamics, 2nd ed. New York: Praeger; 1983, page 82.
4. Institute of Medicine. To err is human: building a better health system. Washington DC: National Acad Press; 1999.
5. Gallagher TH et al. Patients' and physicians' attitudes regarding the disclosure of medical errors. JAMA. 2003;289:1001.

There Are None Whom We Cannot Harm

There are some patients whom we cannot help; there are none whom we cannot harm.

American physician and educator Arthur L. Bloomfield (1888–1962) [1]

Arthur L. Bloomfield, infectious disease specialist and professor of medicine at Stanford University School of Medicine (see Fig. 9.5), was a leader in America's study of penicillin and was one of the first physicians to use the drug to treat bacterial endocarditis. He was considered an outstanding diagnostician and bedside teacher [2].

This maxim, known to most practicing physicians today, brings to mind a true story. The patient was an 18-year-old female college freshman with a past history of depression who presented to the emergency room and was subsequently admitted with diagnosis of "viral syndrome with hysterical symptoms." In short, she seemed to have a flu-like syndrome to which she was overreacting.

Fig. 9.5 Stanford University School of Medicine. (This file is licensed under the Creative Commons Attribution-Share Alike 3.0 Unported license.) http://commons.wikimedia.org/wiki/File:Lokey_Building.jpg

She was given an injection of meperidine (Demerol) to control her shaking and tucked in bed. A few hours later the patient became more agitated and a tired intern ordered an injection of haloperidol (Haldol) and the use of physical restraints. About 3 h later, the young woman was found to have a temperature of 107 °F. Despite emergency measures to reduce the high fever, she suffered cardiac arrest and died. The actual cause of death, unrecognized at the time, was probably serotonin syndrome, resulting from a drug interaction between meperidine and the antidepressant the patient was taking when admitted—phenelzine (Nardil).

The girl's father, an attorney and a writer, publicly criticized the care his daughter—Libby Zion—had received. The focus was on sleep-deprived house staff officers making crucial decisions. The outcome was the New York State Bell Commission Report whose recommendations, adopted in 1989, limited resident work hours to 80 h per week or no more than 24 consecutive hours [3].

The intern involved in the 1984 episode, probably now a seasoned physician practicing somewhere, tried at the time to help the patient with an injection—actually two injections—to reduce agitation. As it turns out, probably the best course would have been to use no sedative drugs at all.

Scottish physician Sir Robert Hutchison (1871–1960) once wrote: "From inability to let well alone; from too much zeal for the new and contempt for what is old; from putting knowledge before wisdom, science before art, and cleverness before common sense; from treating patients as cases; and from making the cure of the disease more grievous than the endurance of the same, Good Lord, deliver us" [4].

1. Bloomfield AL (attributed). Quoted in: Cuervo LG. The road to health care. BMJ. 2004;329:1.
2. Cox AJ. Arthur L. Bloomfield. West J Med. 1962;97:191.
3. Lerner BH. A case that shook medicine. The Washington Post. November 28, 2006.
4. Hutchison R. Modern treatment (letter). Brit Med J. 1953;1:671.

When Ethical Values Collide

Once a doctor subordinates the claims of an individual patient under his care to the abstract claims of society in general, or the hypothetical claims of some possible alternate patient, he has sold the pass.

American theologian Willard L. Sperry (1882–1954) [1]

Sometimes the issues we physicians face are best seen through the eyes of someone outside medicine, a person whose world view is of spirituality and human relationships, and not of arteries, muscles, bones, and bacteria. One of these was Doctor Willard L. Sperry, Dean of the Theological School of Harvard University, Bartlet Professor of Sacred Rhetoric of the Andover Foundation, and author of the book *The Ethical Basis of Medical Practice* [1].

A classic example of conflict in ethical values of beneficence and non-maleficence arises when the well-being of the individual clashes with the greater good of society. Allied Forces generals faced this sort of dilemma in 1944 as they sent young men to die on the beaches of Normandy to achieve the goal of liberating the continent of Europe. Who gets to own land is faced by government in instances in which an individual's property is seized by *eminent domain* in order, for example, to expand a university; just such a case occurred recently near where I live in Virginia as the court ruled that "the Norfolk Redevelopment and Housing Authority did not have the right to condemn a nearby apartment building for Old Dominion University's expansion" [2].

In medicine, however, we believe we have guidance. The Hippocratic oath includes the words "I will prescribe regimens for the good of my patients according to my ability and my judgment and never do harm to anyone," and is silent on the issue of general societal benefit.

But then we face reality, often in the realm of scarce resources. In a disaster situation, optimum resource use is the basis of triage; the guiding principle is likely to be that the resources go to the most salvageable patients, at the expense of those least likely to survive.

In a historic public health decision in the early 1900s, an Irish immigrant cook named Mary Mallon, aka Typhoid Mary, was found to have infected scores of persons with *Salmonella typhi*. She was banished to live in isolation for 3 years on a small island in New York City's East River, a clear example of the claims of society trumping those of the individual.

What about current medical practice? There is a shortage of flu vaccine, and you have a few doses. Your patient, a machinist who is the sole support of a young family, requests a dose of the vaccine. "Doctor, I can't afford to miss work with the flu." Do you give him the vaccine, or refuse because someone older or sicker might need it at some later time? Would you break confidentiality about a patient—such as one with mental illness—if you suspected the patient might be a danger to others? Would you end life-sustaining, and almost surely futile, care in a dying patient if the hospital or insurance company points out the "waste" of resources involved?

The primary ethical value of the Mayo Clinic, formed in 1889, is that the needs of the patient come first [3]. The American Medical Association Code of Medical Ethics echoes this message [4]. One of the physician's chief roles, along with being a diagnostician and healer, is to be on the side of the individual patient—countering societal or even theoretical forces that may compromise care of the ill or injured individual. And yet, what about the needs of other, faceless patients competing for scarce care?

1. Sperry WL. The ethical basis of medical practice. London: Cassell and Company; 1951.
2. Reilly C. High court rules against ODU in eminent domain case. Virginia Pilot. September 13, 2013.
3. Fye WB et al. The origins and evolution of the Mayo Clinic from 1864 to 1939: A Minnesota family practice that becomes an international "medical Mecca." Bull Hist Med. 2010;84:323.
4. The American Medical Association Code of Medical Ethics. Available at: http://www.ama-assn.org/ama/pub/physician-resources/medical-ethics/code-medical-ethics/opinion10015.page?.

Textbooks and the Avoidance of Standardization

The Egyptians made a fatal mistake. They wrote textbooks, the hermetic books. They made another and more serious mistake, and that was to believe that the textbooks were correct. So they forbid physicians, at peril of their lives, to depart in any way from the treatment prescribed in the hermetic books. It was a remarkable experiment …. The experiment demonstrated that standardization can halt advance but it does not in any way hinder retrogression.

Canadian neurosurgeon Wilder L. Penfield (1861–1976) [1]

I first came to know Wilder Penfield as the author of *The Torch*, a fictionalized account of the life of Hippocrates on the island of Kos [2]. But the author was also an internationally renowned neurosurgeon, a pioneer in mapping the cortices of the brain, and recipient of the 1960 Lister Medal to acknowledge his advancement of surgical science [3]. Penfield's comments quoted above raise issues pertinent today and for medicine's future (see Fig. 9.6).

Fig. 9.6 Wilder Penfield.
http://www.springerimages.
com/Images/Biomedicine/
1-10.1007_978-1-4419-1223-
7_3-10

In Chap. 6 I discussed prescribed treatments, aka clinical guidelines, including the curious ways some are developed, the potential for bias, the inconsistencies among recommendations emanating from various sources, and the stifling effect they can have on innovation. Here I focus on textbooks, or more specifically, medical reference books.

I am the editor of a dozen such books, including one of the two "big books" in the specialty of Family Medicine [4]. Fundamentally, a medical reference book is created as an editor and publisher agree to compile and print the chapters submitted by a number of contributors. If the chapter author makes an error or omits important current information, and if the editor fails to notice the misinformation, it goes into print—for the next 4 or 5 years, until corrected in the next edition [5, 6]. And, occasionally, misinformation is carried forward from one edition to the next.

Add to this the fact that chapters in medical books are written a year or more before the final book is published and sold. Thus some textbook/reference book information is out of date on the day the first copy comes off the press. Granted, most of what is in basic medical books is correct and useful but, as they were traditionally printed and distributed, there has always been some room for questioning.

To help avoid ossifying standardization of care and the perpetuation of misinformation, modern medical informatics has come to the rescue. The geeky term Web 2.0 has been declared the millionth word in the English language [7]. Web 2.0 describes websites, such as social media, which go beyond static pages and allow dialogue between those who produce data and those who read it. Interactive "book" publishing is a version of this. For example, today, as I write this book, the seventh edition of *Family Medicine: Principles and Practice* is being compiled, and will chiefly be distributed online, allowing instant feedback to authors regarding new information or even possible errors, and permit prompt updating or correction, without waiting for the eighth edition to be printed on paper.

Today's informatics may well save us from the stifling standardization and occasional error perpetuation of yesterday's textbooks.

1. Penfield W. The second career. Boston: Little, Brown; 1963.
2. Penfield W. The torch. Boston: Little, Brown, 1960.
3. Lister medal. Ann R Coll Surg Engl. 1961;28:15.
4. Taylor RB, ed. Family medicine: principles and practice, ed 6. New York: Springer; 2003.
5. Baker CL. The quality of medical textbooks: bladder cancer diagnosis as a case study. J Urol. 1999;161:223.
6. Anthony D. The treatment of decubitus ulcers: a century of misinformation in the textbooks. J Adv Nurs. 1996;24:309.
7. Singh A. Millionth word in the English language—Web 2.0. The Telegraph. June 10, 2009.

In Praise of Restraint in Health Care

> The great secret of doctors, known only to their wives, but still hidden from the public, is that most things get better by themselves.

> American physician and educator Lewis Thomas (1913–1993) [1]

Lewis Thomas was not only an outstanding physician, researcher, and award-winning author. He also possessed administrative abilities, serving as dean of both the New York University School of Medicine and Yale Medical School, before becoming president of the Memorial Sloan-Kettering Institute.

Following the words above, he goes on to write: "Most things, in fact, are better by morning. Obviously, it is a great time-saver and money-saver for the physician's family that anxiety about disease is not handled as though it were the disease itself; there is perhaps greater willingness to accept anxiety as a natural, often transient, phenomenon. And certainly there is much less ambition to deploy the full technology of medicine as a corrective for the human condition" [1]. If only all patients

Fig. 9.7 A Ulysses mosaic in the Bardo National Museum, Tunisia (Public domain). http://commons.wikimedia.org/wiki/Category:Odysseus#mediaviewer/File:Tunisia-4727_-_Ulysses.jpg

could enjoy the same restraint experienced by doctors' families. Yes, the physician's family sometimes receives suboptimal care—the shoemaker's children going barefoot. But more often it is preferable to exercise caution in undertaking extensive diagnostic testing and perhaps invasive procedures for the sundry aches and pains we all suffer.

This brings me to the Ulysses syndrome, described by Rang, and experienced by many unsuspecting patients [2]. It begins with an abnormal test result, generally one that is unexpected, in an apparently healthy person. (There is, of course, a strong statistical probability that if 20 chemical tests are done, one value will be outside the reference range, given that "abnormal" laboratory reference values are described as the top and bottom 2.5 % of findings on the bell curve.) Once an abnormal value is found, such as a mildly elevated alkaline phosphate (ALP) level or a curious shadow on a chest film, then the patient is launched on an odyssey of testing to detect some elusive, but probably nonexistent, disease cause (see Fig. 9.7).

Mass screenings for breast cancer or coronary artery disease have caused many cases of the Ulysses syndrome. The prototype for the invasive quest for treatable disease begins when an older man is found to have an elevated prostate-specific antigen (PSA) test value. The risks for unnecessary surgery are high and, even so, biopsy procedures are often recommended to guide what may be questionable decisions.

It is no wonder that wise physicians, cognizant of the risks of the diagnostic odyssey, are reluctant to subject themselves and their families to the over-testing and overtreatment that can often be avoided by seeing what the morning brings.

1. Thomas L. Aspects of biomedical science policy. Washington DC: Institute of Medicine National Academy of Sciences; 1972, page 4.
2. Rang M. The Ulysses syndrome. Can Med Assn J. 1972;106:112.

Sometimes Depression and Loneliness

It is by now one of the world's most poorly kept secrets that anxiety, depression, loneliness, and burnout are major factors in the lives of many doctors.

America physician and author David Hilfiker (1945–) [1]

David Hilfiker is a family physician who practiced for 7 years in rural Minnesota. Then, exhausted and burned out, he took a year-long sabbatical in Finland (his wife's home), followed by a move to Washington DC. Here he devoted the next 10 years to serving the poor in a faith-based clinic far from the city's affluent neighborhoods. In 1984, despite a warning from editor Dr. Arnold Relman that he risked "serious damage to his career," Hilfiker published an article in the *New England Journal of Medicine* titled "Facing Our Mistakes" [2]. In this article, he detailed some of his own errors, such as once aborting a healthy fetus after having misdiagnosed it as a fetal demise. This article was followed by the 1985 publication of his book *Healing the Wounds: A Physician Looks at his Work* [1].

Hilfiker has been, and is, very open about his professional failings and his life, including his feelings about his medical practice. In his website, he describes past feelings of being "depressed, unhappy, and feeling little self-worth" [3].

I thought at length about the quotation above. In a book dedicated to wise, and often inspiring, thoughts, should I include a message about anxiety, depression, and loneliness? Yet, what physician has not felt these emotions at times when things go wrong?

In the literature there are two fictional, yet classic, tales of physician despair. The first is Sinclair Lewis' Doctor Martin Arrowsmith who, frustrated with rural practice, concludes: "I shall never practice medicine again.... I'm no good...I'm through. I'll go get a lab job." He does so, and enjoys professional success before undergoing yet another career upheaval [4]. In George Eliot's *Middlemarch*, the protagonist Doctor Tertius Lydgate, after some professional and personal misjudgments, declares: "My practice and my reputation are utterly damned—I can see that." Lydgate, with his family, leaves village practice and moves to London to build a practice catering to the wealthy [5].

Although many—perhaps most—of us have experienced all the feelings described above, we have also learned that, like the flu, they tend to pass. To use the words of Lewis Thomas, above, they are "better in the morning," or perhaps after caring for a newborn or hearing words of gratitude from a patient.

Here is another way to combat professional ennui: Find and reread the essay you wrote when an applicant to medical school, the one titled "Why I Want to Be a Physician." After all, despite the occasional frustration, misadventure, or failure, medicine remains, in the words of Lord Lister (1827–1912), "a noble and holy calling" [6], one that we physicians are privileged to serve.

Hilfiker is an example of nobility and courage. As I write this, he is, at age 68, the author of a web page detailing his long struggle with depression culminating in a now-challenged diagnosis of some type of "mild cognitive impairment." His autobiography, which I recommend to you, describes his journey, including an earlier diagnosis of Alzheimer disease [3]. As a good physician, Hilfiker is still serving others.

1. Hilfiker D. Healing the wounds: a physician looks at his work. New York: Pantheon; 1985, page 11.
2. Hilfiker D. Facing our mistakes. N Engl J Med. 1984;310:118.
3. Hilfiker D. Watching the lights go out: an autobiography. Available at: http://davidhilfiker.blogspot.com/2013/10/letting-go-of-alzheimers.html.
4. Lewis S. Arrowsmith. New York: New American Library; 1961, page 156.
5. Eliot G. Middlemarch. First published in 1874. Available from: New York: Dutton; 1965, page 278.
6. Lister J. Address to the University of Edinburgh. August, 1876. In: McDonald. Oxford dictionary of medical quotations. New York: Oxford University Press; 2004, page 61.

Swimming in the Waters of Uncertainty

> In medicine, uncertainty is the water we swim in.
>
> American physician and author Lisa Sanders (1956–) [1]

Medical author and journalist, Lisa Saunders is an attending physician at Yale-New Haven Hospital, the institutional inspiration for the hospital depicted in the television series *House MD*. She is a New York Times columnist and author of several books about medicine, including the source of the quote above. She is the youngest "giant" described in the book, and her work cited is the most recently

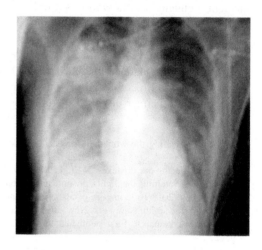

Fig. 9.8 Chest X-ray of a patient with SARS. http://commons.wikimedia.org/wiki/File:SARS_xray.jpg

published. I especially like the metaphorical aphorism. Of all the sayings in this book, hers is one that is likely to stick with you.

Every diagnosis has a hint of doubt. Does the patient really have appendicitis? Or, as is the case in approximately 10 % of nonobese subjects in one study, a normal appendix is found at surgery [2]. Every therapy has a threat of unpredictability. If I give penicillin to a patient with a strep throat, one or more of several things may occur: The patient may improve, which we all want to happen; the patient may develop an allergic reaction to penicillin; diarrhea may occur after a few days; or nothing at all may change with the pharyngitis persisting.

On an international level, the emergence of the severe acute respiratory syndrome (SARS) was certainly a time of uncertainty. Wenzel et al. describe how, in retrospect, the first human SARS victim was a traveler from an agricultural area of China in November 2002. The world began to suspect a problem 3 months later, when seven persons staying on the same floor in the same hotel in Hong Kong contracted the virus and carried it home to five different countries. By April, 2003 SARS, with a death rate of nearly 5 %, had been reported in 27 countries, as public health authorities, swimming in a sea of uncertainty, urgently sought to find the cause and a way to stop its spread [3] (see Fig. 9.8).

Just to extend the "swimming" metaphor a bit. The "water" also contains microorganisms, such as the human immunodeficiency virus Ebola virus, or *Mycobacterium tuberculosis*, to which health providers are not immune. Enter the water with care.

1. Sanders L. Every patient tells a story: medical mysteries and the art of diagnosis. New York: Harmony Books; 2010.
2. Kutasy B. Increased incidence of negative appendectomy in childhood obesity. Ped Surg Int. 2010;26:959.
3. Wenzel RP et al. Managing SARS amidst uncertainty. N Engl J Med. 2003;348:1947.

Chapter 10
Current Issues and Future Practice

In 1900 the leading cause of death in the USA was pneumonia/influenza, with tuberculosis a close second. In that era, a diagnosis of diabetes mellitus could be a death sentence. Today the leading causes of death are heart disease and cancer. Pneumonia/influenza is now number eight on the list, while tuberculosis does not make the "top 10" [1].

Much has changed since 1900: Frederick Banting (1891–1941) and Charles Best (1899–1978) in Canada first tested insulin therapy in a human in 1922. We finally developed effective antimicrobials in the 1930s and 1940s. We recognized the presence of a new disease—acquired immunodeficiency syndrome (AIDS)—in the early 1980s, and the Human Genome Project was declared complete in 2003 (at a cost of $2 billion). We now have virtual colonoscopy, transcranial magnetic stimulation to treat migraine headaches, and "smart" insulin pumps for diabetic patients (see Fig. 10.1). All this just sets the stage for what is to come.

Gene therapy is becoming a reality, with techniques that piggyback a functional gene to a virus that delivers it to a defective cell's nucleus [2]. For more than a decade scientists have worked to develop gene therapy for cystic fibrosis (CF) [3]. We are developing drugs, such as MK3475 (Merck & Co.), that can help empower

© Springer Science+Business Media New York 2015
R.B. Taylor, *On the Shoulders of Medicine's Giants*,
DOI 10.1007/978-1-4939-1335-0_10

Fig. 10.1 Smart insulin
pump. http://www.springer
images.com/Images/
MedicineAndPublicHealth/
2-ACE0203-20-018

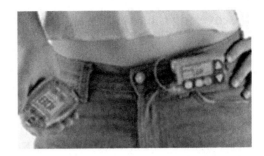

immune cells to combat malignant tumors such as lung cancer and melanoma [4]. Someday the personal physician may use DNA sequencing as part of the patient's routine checkup [5]. The race is on to develop, adapt, and market tomorrow's amazing breakthroughs.

The future, however, is not just about gee-whiz biomedical innovations. As you will see in this chapter, there will also be changes in the domains of language, risk management in hospitals, coordinated team care, the reports of clinical research in medical journals, and how we are permitted to practice our profession. These changes, which will affect both physicians and our patients, may prove just as important as the technologic wonders sure to come. Our response to what happens will determine how history regards the physicians of our era.

1. Leading causes of death. Centers for Disease Control and Prevention. Available at: http://www. cdc.gov/nchs/fastats/lcod.htm.
2. What is gene therapy? Available at: http://ghr.nlm.nih.gov/handbook/therapy/genetherapy.
3. Griesenbach U et al. Progress in gene therapy for cystic fibrosis lung disease. Curr Pharmaceut Design. 2013;18:642.
4. Inman S. Early results show promise for Anit-PD-1 antibody MK-3475 in NSCLC. Available at: http://www.onclive.com/web-exclusives/Early-Results-Show-Promise-for-MK-3475-in-NSCLC.
5. Race to cut whole genome sequencing costs. Genetic Engineering & Biotechnology News. Available at: http://www.genengnews.com/gen-articles/race-to-cut-whole-genome-sequencing-costs/939/.

Discovering the Motions of the Internal Organs

I have been able to hear very plainly the beating of a man's heart …. Who knows, I say, that
it may be possible to discover the motions of the internal organs … by the sound they make.

English scientist and philosopher Robert Hooke (1635–1703) [1]

In the seventeenth century Robert Hooke made contributions in biology,
paleontology, astronomy, and gravitation. He also engaged in lively, and at times
acrimonious, correspondence with Isaac Newton (1642–1727). In this series of
letters we find Newton's famous "standing on the shoulders of giants" image that
inspired the title of this book [2] (see Fig. 10.2).

Fig. 10.2 Robert Hooke.
No contemporary portrait
remains of Hooke; this
impression is based on
contemporary descriptions of
his appearance. http://www.
springerimages.com/
Images/Physics/
1-10.1007_978-1-4419-
5592-0_3

Fig. 10.3 The first wooden stethoscope, built by Laënnec himself from boltable component parts.
http://www.springerimages.com/Images/MedicineAndPublicHealth/1-10.1007_s10840-009-9407-6-1

Hooke lived more than a century before Laënnec's 1816 invention of the monaural stethoscope (see Fig. 10.3), and made his observations, in the style of the time, by direct auscultation, placing his ear against the patient's chest. Hooke was, of course, not the first to hear heart sounds. Hippocrates (460–377 BCE) told of hearing the sounds using the same method as did Hooke two millennia later [3]. But, Hooke may have been the first to articulate the potential benefits of studying the sounds. I find an interesting choice of words in the quote above. Hooke, who studied physics and the movement of celestial bodies, spoke of the "motions" of internal organs, and not of their action or function.

Just as we no longer apply our ears directly to our patients' chests, technology threatens to make the study of heart sounds an anachronism. The paradigm shift was the 1903 invention of a practical electrocardiogram by Dutch physician Willem Einthoven (1860–1927). For those who study heart sounds today, there is the phonocardiogram—the electronic intelligent stethoscope [3]. There are also the echocardiogram, the Holter monitor, cardiac catheterization, computed tomographic X-ray scanning of the heart, and other ways to assess cardiac function. Is the day coming when the motions of the internal organs will all be discerned technologically, with scant human intervention?

1. Hooke R. Quoted in: Leatham A. Auscultation of the heart since Laënnec. Thorax. 1981;36:95.
2. Sette P. Standing on the shoulders of giants—Isaac Newton? Bernard of Chartres? Priscian! (letter). Pharos Alpha Omega Alpha Honor Med Soc. 2012;75:1.
3. Ahlström C. Processing the phonocardiographic signal—methods for the intelligent stethoscope. Linköping: Linköping University; 2006, page 1.

Language in Medicine and Other Endeavors

I feel convinced that there will come a day when physiologists, poets and philosophers will all speak the same language.

French physiologist Claude Bernard (1813–1878) [1]

Claude Bernard, a pioneer of the scientific method, blinded experiments and vivisection, promoted the concept of the *milieu intérieur*, what we now call homeostasis [2]. Before becoming a renowned scientist, however, he was an aspiring playwright, which may help explain his dream of universal interdisciplinary communication [3] (see Fig. 10.4).

Will the day ever come when scientists, healers, and those who think lofty thoughts will communicate without misunderstanding, let alone speak the same language? Physiologists, like Bernard and other scientists, strive for precision in language; physicians aspire to clarity in speaking with patients, but must always deal with the affective human component; poets and philosophers often employ emotive

Fig. 10.4 Claude Bernard. http://www.springerimages. com/Images/Medicine AndPublicHealth/1-10.1007_ s00120-009-1956-x-2

and sometimes purposefully imprecise language that enhances the writing. Scientists seek evidence, such as the effectiveness of a new drug; physicians apply evidence to treat very diverse individuals whom the drug may help or harm; poets and philosophers seek to expand the world of the mind, and somehow perhaps bring comfort to the suffering using no drugs at all.

My world is medicine, and we physicians have a huge, rich vocabulary that often offers historical insights such as the Ulysses syndrome described earlier, the Pickwickian syndrome, and Lou Gehrig disease. The current 32nd edition of *Dorland's Illustrated Medical Dictionary* contains 2,176 pages [4]. Most of the words are combinations of Latin and Old Greek roots, many of which eventually become familiar: *kardia* means heart; from this root we get cardiology (*logos* meaning study of), phonocardiogram (*phono*, meaning sound), and so forth.

On balance, there are the poets and philosophers. William Carlos Williams (1883–1963) wrote: "It is difficult to get the news from poems/yet men die miserably every day/for lack of what is found there" [5]. Williams was both poet and practicing physician, and probably intended his reader to speculate as to whether he meant (medical) physical death or (philosophical) spiritual death. Today, several prestigious medical journals, including the *Journal of the American Medical Association* (JAMA), present sections on poetry.

Still, in reports of research studies, there is no room for creative writing, as any author submitting a paper with humor or irony will soon learn. Today's scientific papers are laden with often-arcane statistical terms, and presented in turgid prose. As such, many are incomprehensible to ordinary doctors, not to mention poets and philosophers.

So perhaps Bernard's prophesy of one language will some day prove true, but we have a long way to go.

1. Bernard C. Quoted in: Cousins N. The physician in literature. Philadelphia: Saunders; 1982, Introduction page xxiii.
2. Bernard C. Lectures on the phenomena common to animals and plants. Trans. Hoff HE et al. Springfield, IL: Charles C. Thomas; 1974.
3. Wilson DW. Claude Bernard. Popular Science Monthly. 1914;5:567.
4. Dorland's Illustrated Medical Dictionary. Philadelphia: Saunders; 2011.
5. Williams WC. Asphodel, that greenery flower. New York: New Directions Publishers; 1994, page 19.

Hospitals and Harm, Then and Now

It may seem a strange principle to enunciate as the very first requirement in a hospital that it should do the sick no harm.

British nurse Florence Nightingale (1820–1910) [1]

We have previously discussed "do no harm" as the dictum applies to physicians. But what of hospitals—buildings populated by physicians, nurses, laboratory technicians, administrators, and many other professionals, and, of course, patients?

Florence Nightingale, a friend of another pioneering woman, Elizabeth Blackwell, described in Chap. 2, is best remembered for her work during the Crimean War in the military hospital at Scutari, near Istanbul. When she arrived at the hospital in 1854, she found over-crowding, poor sanitation, medication shortages, and rampant infections. She worked to correct the deplorable conditions she had found, reduced the hospital death rate, and, because she often made rounds at

Fig. 10.5 Florence Nightingale. http://www.springerimages.com/Images/MedicineAndPublicHealth/1-10.1007_978-3-642-29993-3_2-6

night, earned the title "The Lady with the Lamp." In her later career she was a leader in nursing education and an advocate of strict cleanliness in hospitals (see Fig. 10.5).

Over the centuries, there have been some iconic hospitals that have both helped and harmed. One of the earliest was the Aesculapian temple on the island of Kos, where the ailing came to be treated by Hippocrates in the fifth century BCE; the ruins of this building are there to be visited today. The oldest existing psychiatric hospital in Europe, opened in 1330, is Bethlem Royal Hospital in London. The dreadful conditions in this hospital for the insane have given us the word "Bedlam," connoting madness [3]. America's first hospital was Pennsylvania Hospital in Philadelphia, which opened its doors in 1756. Until the mid-1800s women having babies in any hospital risked their lives because of "puerperal fever," transmitted by unwashed physician hands, as exposed by Ignaz Semmelweis in 1847.

There was the fictional sanitarium at Davos, Switzerland, in Thomas Mann's 1924 book *The Magic Mountain*, and the very real Trudeau Sanitarium in Saranac Lake, New York, both offering tuberculosis patients the benefits of clean, cool mountain air. Today visitors to Albert Schweitzer's jungle hospital at Lambaréné in West Africa will find inscribed on the outside lamp, "Here, at whatever hour you come, you will find light and help and human kindness."

Currently American hospitals are undergoing a transition. In my early practice days, a patient with an acute back strain might be admitted for 10 days of bed rest with Buck's extension traction. No more. Hospitals have become large intensive care units. Patients are discharged "quicker and sicker," perhaps for the best because of the rising threat of hospital-acquired infections. Methicillin-resistant *Staphylococcus aureus* (MRSA) has become a widespread institutional problem. An emerging concern is carbapenem-resistant *Klebsiella* and *Enterobacteriaceae*, and multiple-drug-resistant tuberculosis reminds us that old plagues can reemerge. Our exuberant overuse of antibiotics, coupled with an anemic new drug pipeline, raises the spectre of a "post-antibiotic era," in which our most esteemed hospitals risk becoming elegant, but potentially harmful places to be [4].

1. Nightingale F. Preface. Notes on hospitals. London: Longman; 1859.
2. Cook ET. The life of Florence Nightingale: 1862–1910. London: MacMillan; 1914.
3. Andrews J et al. The history of Bethlem. London: Routledge; 1997.
4. Kuehn BM. Nightmare bacteria on the rise in US hospitals, long-term care facilities. JAMA. 2013;309:1573.

Teamwork and Tomorrow's Practice

No one is big enough to be independent.

British-American physician William Worrall Mayo (1819–1911) [1]

According to the institutional website, "Teamwork—the sharing of diverse skills for a common good—is the essence of Mayo Clinic" [1]. William Worrall Mayo, father of the brothers Mayo and the founder of the private practice in Rochester, Minnesota, that later became the Mayo Clinic, instilled in his sons the importance of collaboration by a variety of health professionals (see Fig. 10.6). As an advocate of team care in the late nineteenth century, the senior Doctor Mayo was ahead of his time [2] (see Fig. 10.7).

A century ago, the typical American physician was a solo practitioner who was an entrepreneur and an autocratic decision-maker. Physicians grew more collaborative as specialties began to emerge—the first specialty, ophthalmology, was established in 1916—and as practice partnerships were formed.

Fig. 10.6 William Worrall Mayo. http://www.springer images.com/Images/ MedicineAndPublic Health/1-10.1007_s00105- 003-0675-2-5

A major upheaval occurred with World War II, which uprooted thousands of established generalists who were sent to war. When discharged, many took advantage of the GI Bill of Rights, which supported specialty training. This helped increase the specialization of many who had been general practitioners before the war.

Many became surgeons, using skills learned on battlefields. They formed group practices, often quite large in size. Eventually persons with more focused skills—advance practice nurses, task-specific physician assistants such as those who assist at surgery, and specially trained technicians such as those who apply casts in orthopedic practices—were needed to provide increasingly complex care.

Today, in addition to the physician, the team line-up will include a number of persons whose focused knowledge of some aspect of pharmacology, anesthesia, nursing care, physical therapy, social services, and other aspects of management complements that of the physician [4]. Increasingly—for better or worse—the physician's role may evolve from provider to supervisor of care.

1. Mayo WW. Quoted in: The bonds of brotherhood, teamwork and group practice. Mayo Clinic/ patient care. Available at: http://www.mayoclinic.org/tradition-heritage/brothers-practice.html.
2. Clapesattle H. The doctors Mayo. Minneapolis: University of Minnesota Press; 1975.
3. Starfield B. Contribution of primary care to health systems and health. Milbank Quarterly. 2005;83:457.
4. King H. Teamwork improvement in health care. In: Developing and enhancing teamwork in organizations: Salas E et al., eds. New York: Wiley; 2013, page 298.

Contract Practice in the Twenty-First Century

> Contract practice, in which a doctor treated the members of a friendly society for an agreed sum, or formed his own penny-a-week club among his flock, was already in existence in the early Victorian years. To physicians of the old school this form of practice was an anathema. It was undignified. It was mere trading. It was making the doctor a hireling of Buffaloes and Foresters and secret societies.
>
> English author Ernest Sackville (E.S.) Turner (1909–2006) [1]

E.S. Turner was a writer who provided a historical insight into health care delivery. I include him here because of his vivid writing about contract practice. His words prompted me to search for the meaning of *Buffaloes*, capital B. In the context used, Turner refers to the *Royal Antediluvian Order of Buffaloes*, a large British fraternal order founded in 1822. *Foresters* denotes the *Ancient Order of Foresters*, a "friendly society" founded in Britain in 1834.

The intriguing phrase for me, however, refers to the physician forming "his own penny-a-week club among his flock." This brings me to concierge medical practice, a model of health care delivery that, at least in North America, is increasing in popularity. Sometimes called "boutique," retainer, or membership practice, concierge medicine involves payment by the patient of a monthly or yearly fee to the physician. In return the patient is promised unhurried office visits and access by telephone or e-mail at any time. Patients in concierge practices can call the physician's cell phone, unheard of in larger, traditional practices. Most doctors offering concierge medicine are primary care physicians—general internists or family physicians—and their practices involve approximately a third the number of patients found in a non-concierge practice [2]. In one study, retainer physicians had smaller patient panels than traditional physicians: mean 898 vs. 2,303 patients [3].

Physicians are attracted to concierge medicine by avoidance of many bureaucratic regulations and the ability to see patients at a relaxed pace. If you are a primary care physician seeing 20-plus patients per day, consider a practice involving, for example, eight office patients a day. No waiting for delayed or contested "reimbursement" by an insurance company or government agency. As the US Affordable Care Act rolls out, can we expect to see more and more physicians start membership practices? [4].

Detractors criticize the concierge model, claiming that it is only for the healthy and wealthy, forcing other physicians to care for the poor. As to being "only for the healthy," concierge medicine may actually be best suited to those with chronic diseases, such as diabetes [5]. As to the "wealthy" criticism, many concierge doctors share the obligation to care for the needy. In the Alexander et al. study cited above, 5 % of patients in concierge practices were Medicaid patients. The authors report: "Most retainer physicians (84 %) provide charity care and many continue to see some patients (mean. 17 % of the panel) who do not pay retainer fees" [3].

According to an August 2013 report, concierge medicine offers more satisfying work, less overhead, and more personal care for patients. This study by *The Physicians Foundation* showed that "almost 1 in 10 practitioners are considering converting to a private or concierge model in the next three years" [6].

Although one author writes that concierge medicine began in Seattle in the mid-1990s [5], the "penny-a-week" model of health care dates, according to Turner, to Victorian times [1]. Only the cost per week and the sophistication of services offered have changed.

1. Turner ES. Call the doctor: a social history of medical men. New York: St. Martin's Press; 1959.
2. French MT et al. Is the United States ready to embrace concierge medicine? Population Health Med. 2010;13:177.
3. Alexander GC et al. Physicians in retainer ("concierge") practice: a national survey of physician, patient and practice characteristics. J Gen Intern Med. 2005;20:1079.
4. Page L. The rise and further rise of concierge medicine. BMJ. 2013;347:f6465.
5. Schyberg JP. Nontraditional or noncentralized models of diabetes care: boutique medicine. J Fam Pract. 2011;60:S19.
6. How concierge medicine benefits care providers. Available at: http://www.aapp.org/blog/page/2/.

Doctors, Merchants, and Giants

> It is because we have begun to act like merchants, and in many instances to observe the same hours, that the public expects us to be regulated by the same restraints.
>
> American physician and author John L. McClenahan (1915–2008) [1]

John L. McClenahan, Philadelphia radiologist and editor of the journal *Transactions and Studies,* voiced concern in 1961 about the tendency among physicians to seek "banker's hours." Writing four decades later regarding surgical residents, Sanfey observed [2]:

> Today's trainees have different values and demand a more balanced lifestyle than those who believed the only thing wrong with every-other-night call was that you missed half the good cases.

Young physicians now actually want to be married, get to know their children, and have a life outside medicine. This heresy defies tradition. The word "intern" can be traced back to meaning one who is a resident dwelling within a school or, in the case of medicine, within the hospital. The medical or surgical resident "resided" in the hospital. Many hospitals refused to accept married residents. Physicians were expected to be celibate "iron men," and to carry this mentality into practice. In *Horse and Buggy Doctor,* rural physician Arthur E. Hertzler (1870–1946) described an early 20th doctor: "He lived for his patients like all country doctors. He had few social contacts. He joined the lodge but seldom attended. He fought mud and snow until he was exhausted—it was the same from year's end to year's end" [3].

Much has changed in recent years in regard to the doctor's practice model and hours on the job. Resident work hours have been capped by legislation, imparting a "shift work" mentality. As a result, residents can leave their patients in the morning when their duty hours are up, not sure what will happen after they depart. Lifestyle has become an important issue for aspiring physicians. Some young doctors are seeking employment "three days a week, no night call, please." There are persons with M.D. degrees today who, frankly, see medical practice as a means to support their true loves—music, skiing, or travel.

Managed care organizations and large group practices must share the blame for the "merchant" mentality. We have moved operational decisions from physicians to administrators. Who will work what days of the week and what hours of the day are often dictated to doctors, not decided by them. And the employment of hospitalists, laborists, and nocturnalists has converted health care in many institutions into the hourly duties that residents are learning in training.

The result has been a loss of professional prerogatives and increased regulation in many spheres. In large groups, administrators may even tell doctors how many minutes they can spend with each patient and dictate the menu of drugs that can be prescribed. Federal law now governs how much we can bill our Medicare patients, and what we will be paid. It makes the time-honored practice of extending professional courtesy to colleagues risky, with the threat of violating anti-kickback statutes or committing "fraud" [4]. We are told that we can or cannot obtain certain tests or even admit a sick patient to the hospital. In some states there is legislation regarding what we can tell pregnant patients.

Have we, in seeking a lifestyle similar to that enjoyed by some of our patients, sold our professional souls? Can we get back some of what we had before we began to act like merchants and seek their hours? Can physicians recover the prestige that we earned in centuries past? Will there be giants in the current generation of physicians? I will return to this important issue in the last essay of the book.

1. McClenahan JL. Editorial. X-ray technician. 1961;32:498.
2. Sanfey H. General surgery training crisis in America. Brit J Surg. 2002;89:132.
3. Hertzler A. Horse and buggy doctor. New York: Harper; 1938, page 36.

Primary Care and Equitable Distribution of Health

> The evidence … shows that primary care (in contrast to specialty care) is associated with a more equitable distribution of health in populations, a finding that holds in both cross-national and within-national studies.
>
> American physician-researcher Barbara Starfield (1932–2011) [1]

Society—and I am taking a worldview here—does not need more pediatric dermatologists, plastic surgeons, or sports medicine specialists. The world needs more primary care physicians: that is, family physicians, general internists, and general pediatricians. It needs doctors who will provide first contact, comprehensive, and continuing care. It needs doctors who will, when necessary, speak up to help their patients get scarce resources when needed. It needs doctors who will provide comfort when pharmaceuticals and technology fail. The world needs personal physicians.

There is a desperate shortage of physicians in many countries, especially in Southeast Asia, sub-Saharan Africa, and South America [2]. In these countries, people don't necessarily die of the same diseases we die of here in America. Yes, some low-income country inhabitants die of coronary artery disease and cancer, just as do those in high-income countries. But those in doctor-poor areas are much more

Fig. 10.8 Barbara Starfield. (This file is licensed under the Creative Commons Attribution-Share Alike 3.0 Unported license.) http://commons.wikimedia.org/wiki/Category:Barbara_Starfield#mediaviewer/File:Barbara_Starfield.jpg

likely to die of AIDS, diarrheal disease, malaria, and tuberculosis, diseases that don't top the list of causes of death in high-income countries [3]. These are also diseases that can often be prevented or treated effectively if there were more generalist physicians, nurses, and medication available.

For decades Barbara Starfield described primary care as the key to the best health care to the most people, both in North America and around the world [4]. She was a pediatrician and the head of the Department of Health Policy and Management of the Johns Hopkins Bloomberg School of Public Health. As cofounder and first president of the International Society for Equity in Health, she sought to bring quality health care to all (see Fig. 10.8).

Starfield and her colleagues published a number of papers documenting the value of primary care in the health care system. In one paper, the authors analyzing ten studies of primary care physician supply and health outcomes report: "Primary care physician supply was associated with improved health outcomes, including all-cause, cancer, heart disease, stroke, and infant mortality; low birth weight; life expectancy; and self-rated health …. Pooled results for all-cause mortality suggest that an increase of one primary care physician per 10,000 population was associated with an average mortality reduction of 5.3 %, or 49 per 100,000 per year" [5].

Barbara Starfield was a tireless spokesperson for primary care as the best answer to what ails our health care system, and I heard her speak on this topic several times. I hope that some day our policy makers here and abroad will give some thought to offering the best health care to the most people, and provide the incentives needed for many more medical graduates, including the best and the brightest, to choose primary care careers.

1. Starfield B et al. Contribution of primary care to health systems and health. Milbank Q. 2005;83:457.
2. Bossert TJ et al. Finding affordable health workforce targets in low-income nations. Health Affairs (Millwood). 2010;29:1376.
3. The top ten causes of death by broad income group. World Health Organization. Available at: http://www.who.int/mediacentre/factsheets/fs310.pdf.
4. Starfield B. Primary care: concept, evaluation and policy. New York: Oxford University Press, 1992.
5. Macinko J, Starfield B et al. Quantifying the health benefits of primary care physician supply in the United States. Int J Health Services. 2007;37:111.

The iPatient and the Virtual Physician

> The patient is still at the center, but more as an icon for another entity clothed in binary garments: the "iPatient." Often, emergency room personal have already scanned, tested, and diagnosed, so that interns meet a fully formed iPatient before seeing the real patient.
>
> Physician and author Abraham Verghese (1955–) [1]

An exemplar of a physician with physical examination skills in the Oslerian tradition and author of the best-selling, somewhat autobiographical novel *Cutting for Stone* [2], Verghese writes, "Because the electrocardiogram, magnetic resonance imaging and computed tomographic scan precisely characterize anatomy, the physical exam is too often viewed as redundant." Placing hands on the patient has

Fig. 10.9 Abraham Verghese. This file is licensed under the Creative Commons Attribution 3.0 Unported. http://commons.wikimedia. org/wiki/File:Verghese,_ Abraham,_blurred_2.jpg

metamorphosed to fingers tapping on a computer keyboard [1] (see Fig. 10.9). More recently, physician author Robin Cook has brought us his 2014 novel *Cell*, in which the smartphone—the iDoc—becomes capable of diagnosing and treating disease [3].

Enter the virtual physician. The Wall Street Journal recently printed a story titled "Doctors Move to Webcams," describing virtual doctor visit services for "patients whom they meet online via online video or phone." The vehicles will be laptop webcams or video-enabled tablets and smartphones. Health insurer *WellPoint*, according to the article, plans to offer virtual doctor visits in all of its health insurance plans [4].

What's left to bring patient and physician together in the same room? Only the physical examination remains. A patient can describe symptoms in an e-mail message and transmit an electrocardiogram or blood sugar level by smartphone. Medication can be prescribed electronically. Webcams can provide a few basic clues to physician findings—the appearance of a skin rash or extent of a laceration—but images are imperfect and the repertoire of what can be examined in this manner is limited.

All this emphasizes the importance of teaching physical examination skills to young physicians. First of all, the physical examination can provide the most important clues in diagnosis—the calf tenderness of deep vein thrombosis of the calf, the Virchow node sometimes seen with gastric cancer or borborygmi, the elephantine rumbling of a hyperactive bowel.

There is also the value of the personal bonding with the patient, the opportunity to continue following threads of the patient's story while performing the examination, and the therapeutic touch of the healer, which has healing properties of its own—all missing with the iPatient, the iDoc, and the virtual physician.

1. Verghese A. Culture shock—patient as icon, icon as patient. N Engl J Med. 2008;359:2748.
2. Verghese A. Cutting for stone. New York: Vintage Books; 2010.
3. Cooke R. Cell. New York: Putnam; 2014.
4. Mathews AW. Doctors move to webcams. Wall Street Journal. December 21, 2012, page B1.

Paying Attention to What Is Going on
Around Us from Now On

My father was a surgeon, and a very good one and a very hard-working one. He never got involved in non-clinical aspects of healthcare, and when as a medical student I asked him why, he honestly answered that he was too busy taking care of patients and never thought there would be any problems with "the other stuff."

Well, sorry Dad, while you and your colleagues were taking care of patients—and doing a great job I should add—the suits, some physician, some not, were taking control. It was so incremental that you didn't see it coming, and I understand that. Now 96, my father recently told me, "We should have paid more attention to what was going on around us."

American physician Richard E. Waltman, Tacoma, Washington, USA [1]

Somehow I think it fitting that this, the ultimate quotation in the book, is from an old-fashioned community physician, as reported by his son, a practicing family physician. Richard Waltman's father probably never spoke to a national medical audience, deduced the cause of an epidemic, or invented a new drug. If we take the words above quite literally, he never even held office in a medical organization or met with legislators regarding the healthcare issues of his day. But his story holds a compelling message for our future—and the fate of health care in America.

In his 2011 *Medical Economics* magazine article, Waltman, the son, describes what prompted him to write: "I received an email from our chief executive officer (CEO) indicating that the healthcare system for which I work was going to reduce the workforce by 350 people—as 'phase 1 in dealing with the budgetary crisis.'" [1] Then, in November 2013 the American public learned that UnitedHealth has "dropped thousand of doctors from its networks in recent weeks" [2]. A few weeks later I read, "Insurers are slashing payments to medical practices in many of the plans they sell through the new health-law marketplaces" [3]. As I write this in early 2014, the promise that helped sell the Affordable Care Act—the one that "if you like your doctor, you can keep your doctor"—has proved false.

Physician and author Martin H. Fischer (1879–1962) observed more than a half century ago: "If the medical profession has problems, it is because it has relinquished what it should of held, or done badly what irregulars have done better" [4]. The list of items we have relinquished is long and sad. We have surrendered our right to choose how we spend our workday, to set our fees, to be paid fairly for our work, and to be free from the arbitrary loss of our jobs. We have allowed ourselves to be diminished by being called "providers," and not physicians. The "suits" are working very hard to transform us into civil servants; think of postal workers in white lab coats. One difference is that we lack an effective union. And we seem to lack the will to engage those who threaten the profession.

The game is not lost, however. Consider the adage: "In any system, whoever controls the scarce resource controls the system." The "suits" do not see office patients, they do not make house calls, and they do not perform surgery. We—the physicians—are the "scarce resource" in the health care system. And that system can still be rescued. Perhaps the retainer practice model, led by physicians and eschewing insurance and government health payments, or the patient-centered medical home (PCMH), or direct primary care will provide the needed stimulus. But we cannot look to the "suits," not even doctors wearing black jackets and trousers that match, to save the future of medicine and preserve the service orientation of health care professionals.

It is up to us, our generation of healers, to identify and support contemporary medicine's giants who will assure the future of our noble profession and upon whose shoulders the next generation of physicians may proudly stand.

1. Waltman RE. Who can solve the healthcare crisis? Physicians can—and must! Med Econ. 2011;25:8.
2. Beck M. UnitedHealth culls doctors from plan. Wall Street Jour. 2013; Nov. 16–17:B1.
3. Weaver C et al. Insurers cut doctors' pay under new health plans. Wall Street Jour. 2013;Nov. 22;A1.
4. Fischer MH. Quoted in: Fabing HD et al., ed. Fischerisms: being a sheath of sundry and diverse utterances culled from the lectures of Martin H. Fischer, professor of physiology in the University of Cincinnati: Cincinnati OH: Bernard Smith; 1956, page 76.

Acknowledgements

This book is about the wisdom of men and women who have helped make medicine what it is today. Not many of the persons quoted are still alive today; most are honored in history. I have had the opportunity to meet only a few of the "giants" described: Paul Dudley White, Edmund Pellegrino, and Barbara Starfield. But we know the words of all because they wrote or else, in some instances such as with Theodore E. Woodward, someone recorded their words (or what the writer believed were their words). As to the "giants" of more remote medical history—Hippocrates, Maimonides, Snow, and others—their discoveries and their thoughts set the stage for achievements of those that followed them. All cited in this book have influenced how we teach and practice medicine in the twenty-first century.

The thoughts above are to introduce my belief that we all have "giants" in our own lives, persons whose words, thoughts, and actions have shaped what we are today. For me, there have been many outstanding individuals. Here, in more or less the order in which they entered my life, are some of the physicians who have influenced and inspired me over the years: E. Thomas (Tom) Deutsch, Charles (Chuck) Visokay, Joseph E. (Joe) Scherger, Peter A. Goodwin, Merle Pennington, John Kendall, William (Bill) Toffler, John Saultz, Scott Fields, Daniel J. Ostergaard, Robin Hull, Robert W. (Bob) Bomengen, Ray Friedman, Tom Hoggard, Mary Burry, Ryuki Kassai, Takashi Yamada, Manabu Yoshimura, Michiyasu Yoshiara, Subra Seetharaman, and Richard Colgan.

A heartfelt thank you is due to Coelleda O'Neil, who worked with me on a quarter-century's worth of books, and to Kathy Cacace, Katherine (Kate) Ghezzi, and Janet Foltin of Springer Publishers who encouraged and supported my writing "Medicine's Giants."

Thanks also to my wife, Anita D. Taylor, M.A. Ed., medical educator and author, who, through some 33 books and many published reports, has read every word I ever wrote or edited (and corrected more than a few).

© Springer Science+Business Media New York 2015
R.B. Taylor, *On the Shoulders of Medicine's Giants*,
DOI 10.1007/978-1-4939-1335-0

Bibliography[1]

Ackerknecht EH. History and geography of the most important diseases. New York: Hafner; 1972.

Balint M. The doctor, his patient and the illness. London: Churchill Livingstone; 1957.

Bean RB, Bean WB. Aphorisms by Sir William Osler: New York: Henry Schuman; 1950.

Bollett AJ. Plagues and poxes: the impact of human history on epidemic disease. New York: Demos; 2004.

Bordley J, Harvey A McG. Two centuries of American medicine. Philadelphia: Saunders; 1976.

Breighton P, Breighton G. The man behind the syndrome. Heidelberg: Springer-Verlag; 1986.

Brody H. Stories of sickness. New Haven: Yale University Press; 1987.

Callan JP. The physician: a professional under stress. Norwalk, Connecticut: Appleton-Century-Crofts; 1983.

Cartwright FF. Disease and history: the influence of disease in shaping the great events of history. New York: Crowell; 1972.

Cassell EJ. Doctoring: the nature of primary care. New York: Oxford; 1997.

Colgan R. Advice to the healer: on the art of caring. New York: Springer; 2013.

Collins FS. The language of God. New York: Free Press/Simon and Schuster; 2006.

Dirckx JH. The language of medicine: its evolution, structure, and dynamics, 2nd edition. New York: Praeger; 1983.

Durham RH. Encyclopedia of medical syndromes. New York: Harper and Brothers; 1960.

Ellerin TB, Diaz LA. Evidence-based medicine: 500 clues to diagnosis & treatment. Philadelphia: Lippincott, Williams & Wilkins; 2001.

Evans IH. Brewer's dictionary of phrase and fable. New York: Harper & Row; 1970.

Fabing HJ, Marr R, editors:. Fischerisms, being a sheaf of sundry and diverse utterances culled from the lectures of Martin H. Fischer, professor of Physiology in the University of Cincinnati. Springfield: Illinois: Charles C. Thomas, 1937.

Firkin BG, Whitworth JA. Dictionary of medical eponyms. Park Ridge NJ: Parthenon; 1987.

Fortuine R. The words of medicine: sources, meanings, and delights. Springfield, Illinois: Charles C. Thomas; 2001.

Garland J. The physician and his practice. Boston: Little, Brown and Co.; 1954.

Garrison FH. History of medicine, 4th edition. Philadelphia: Saunders; 1929.

Gordon R. The alarming history of medicine: amusing anecdotes from Hippocrates to heart transplants. New York: St. Martin's Griffin, 1993.

---

[1] This is a list of books recommended for the reader interested in the wisdom of the great physicians of the past, the language of medicine, the process of clinical practice, and how to communicate with patients and with one another.

© Springer Science+Business Media New York 2015
R.B. Taylor, *On the Shoulders of Medicine's Giants*,
DOI 10.1007/978-1-4939-1335-0

Haubrich WS. Medical meanings: a glossary of word origins. Philadelphia: American College of Physicians; 1997.

Huth EJ, Murray TJ. Medicine in quotations: view of health and disease through the ages. Philadelphia: American College of Physicians; 2006.

Inglis B. A history of medicine. New York: World; 1965.

Johnson WM. The true physician: the modern doctor of the old school. New York: Macmillan; 1936.

Johnson S. The ghost map: the story of London's most terrifying epidemic—and how it changed science, cities, and the modern world. New York: Riverhead Books, 2006.

Lindsay JA. Medical axioms, aphorisms, and clinical memoranda. London: H.K. Lewis Co.; 1923.

Lipkin M. The care of patients. New York: Oxford; 1974.

Magalini SI, Scrascia E. Dictionary of medical syndromes, 2nd edition. Philadelphia: Lippincott; 1981.

Maimonides M. Medical aphorisms: treatises 1–5, Bos G, ed. Provo UT: Brigham Young University Press; 2004.

Major RH. Disease and destiny. New York: Appleton-Century; 1936.

Manning PR, DeBakey L. Medicine: preserving the passion, 2nd edition. New York: Springer; 2004.

Martí-Ibáñez F. Men, molds and history. New York: MD Publications; 1958.

Martí-Ibáñez F. A prelude to medical history. New York: MD Publications; 1961.

Mayo CH, Mayo WJ. Aphorisms of Dr. Charles Horace Mayo and Dr. William James Mayo. Willius FA, editor. Rochester MN: Mayo Foundation for Medical Education and Research; 1988.

McDonald P. Oxford dictionary of medical quotations. New York: Oxford University Press; 2004.

Meador CK. A little book of doctors' rules II. Philadelphia: Hanley & Belfus; 1999.

Meyers MA. Happy accidents. New York: Arcade Books; 2007.

Oldstone MBA. Viruses, plagues and history. New York: Oxford University Press; 1998.

Osler W. Aequanimitas with other addresses. Philadelphia: Blakiston; 1906.

Payer L. Medicine & culture: varieties of treatment in the United States, England, West Germany, and France. New York: Henry Holt; 1988.

Pellegrino ED. Humanism and the physician. Knoxville TN: University of Tennessee Press; 1979.

Penfield W. The torch. Boston: Little, Brown and Co.; 1960.

Porter R. The greatest benefit to mankind. New York: Norton; 1997.

Rapport S, Wright H. Great adventures in medicine. New York: Dial Press; 1952.

Reveno WS. Medical maxims. Springfield IL: Charles C. Thomas; 1951.

Reynolds R, Stone J, editors. On doctoring. New York: Simon & Schuster; 1991.

Ross JJ. Shakespeare's tremor and Orwell's cough. New York: St. Martin's Press; 2012.

Sebastian A. The dictionary of the history of medicine. New York: Parthenon; 1999.

Sherman IW. The power of plagues. Washington, DC: ASM Press; 2006.

Shryock RH. Medicine and society in America: 1660–1860. Ithaca, New York: Cornell University Press; 1960.

Silverman ME, Murray TJ, Bryan CS. The quotable Osler. Philadelphia: American College of Physicians; 2003.

Starr P. The social transformation of American medicine. New York: Basic Books; 1982.

Strauss MB. Familiar medical quotations. Boston: Little, Brown; 1968.

Taylor RB. Medical wisdom and doctoring: the art of 21st century medicine. New York: Springer; 2010.

Taylor RB. White coat tales: medicine's heroes, heritage and misadventures. New York: Springer; 2008.

Weiss AB. Medical odysseys: the different and sometimes unexpected pathways to 20th century medical discoveries. New Brunswick, New Jersey: Rutgers University Press; 1991.

Index

© Springer Science+Business Media New York 2015
R.B. Taylor, *On the Shoulders of Medicine's Giants*,
DOI 10.1007/978-1-4939-1335-0

CPSIA information can be obtained
at www.ICGtesting.com
Printed in the USA
LVOW02*1510111115

462079LV00005B/12/P